Neuroendocrinology of Leptin

Frontiers of Hormone Research

Vol. 26

Series Editor *Ashley B. Grossman*, London

KARGER

Basel · Freiburg · Paris · London · New York · New Delhi · Bangkok · Singapore · Tokyo · Sydney

Neuroendocrinology of Leptin

Volume Editor *Ehud Ur*, Halifax

12 figures and 2 tables, 2000

KARGER

Basel · Freiburg · Paris · London · New York ·
New Delhi · Bangkok · Singapore · Tokyo · Sydney

Ehud Ur, MD, FRCPC
Division of Endocrinology and Metabolism
Dalhousie University
Queen Elizabeth II Health Sciences Centre
Halifax, Nova Scotia, Canada

Library of Congress Cataloging-in-Publication Data

Neuroendocrinology of leptin / volume editor, Ehud Ur.
p. cm. – (Frontiers of hormone research; v. 26)
Includes bibliographical references and index.
ISBN 3–8055–6921–1 (hard cover: alk. paper)
1. Leptin. 2. Neuroendocrinology. I. Ur, Ehud. II. Series.
[DNLM: 1. Neurosecretory Systems – physiology. 2. Proteins – metabolism. W1 FR946F v.26 2000]
QP572.L48N48 2000
612.8′2–dc21
DNLM/DLC
for Library of Congress 99-40628
CIP

Drug Dosage. The authors and the publisher have exerted every effort to ensure that drug selection and dosage set forth in this text are in accord with current recommendations and practice at the time of publication. However, in view of ongoing research, changes in government regulations, and the constant flow of information relating to drug therapy and drug reactions, the reader is urged to check the package insert for each drug for any change in indications and dosage and for added warnings and precautions. This is particularly important when the recommended agent is a new and/or infrequently employed drug.

www.karger.com
Printed in Switzerland on acid-free paper by Reinhardt Druck, Basel
ISSN 0301–3073
ISBN 3–8055–6921–1

Contents

Preface

The discovery of the hormone leptin has become one of the most fascinating discoveries in endocrinology in the 1990s, and the exponential increase in publications on the subject seems set to continue well into the next millennium. Even if the peptide itself does not prove to be of widespread clinical utility, its discovery has led to a massive increase in research into the fundamentals of human obesity, and has completely opened up an area which had previously progressed little over several decades. Several overviews have recently appeared on the subject of leptin and related peptides, but this is the first which concentrates on the neuroendocrine interactions of leptin, in addition to its effect on the control of appetite. Dr. Ur has put together an impressive and wide-ranging series of experts within the field, their reviews emphasising the unique fascination of this novel peptide and its ever-increasing range of pleiotropic activities. In particular, many neuroendocrinologists will welcome the review of the possible involvement of leptin in the induction of puberty, while the recent discovery of leptin in the brain opens up a further field on active research. I am sure that this volume will have much to offer basic scientists and clinicians alike who have been intrigued by the activity of leptin, and who wish to understand the current state of play of research.

Ashley B. Grossman
London, UK

Introduction

Obesity, defined as an increase in body fat, is a growing problem in Western societies. Mean Body Mass Indices (BMI; weight[kg]/height[m]^2) have been steadily increasing in North America and in Western Europe over the last four decades. In the United States over 30% of the population are obese as defined by a BMI greater than 27 kg/m^2. Obesity is associated with increases in hypertension, non-insulin-dependent diabetes, hyperlipidemia, ischemic heart disease and a number of other conditions. The problem of obesity is easily formulated; obesity is the result of a positive energy balance, which comes about as a consequence of an increased ratio of caloric intake to energy expenditure. However, its origins are poorly understood. The heritability of obesity has been estimated to be around 30%. A number of rodent models exist for genetic obesity. The genes responsible for several of these genotypes have been cloned with the result that a novel physiological system based on leptin and its receptors has been identified. Human homologues have also been identified for some of these mutations. Following its isolation in 1994 studies on leptin have generated over 2,000 medline citations and leptin is now undergoing phase II therapeutic efficacy trials in the US, Canada and Europe in subjects with both obesity and type 2 diabetes. The evidence indicates that leptin plays a crucial role in the regulation of body fat levels by coordinating metabolism, feeding behaviour, energy balance and neuroendocrine response. Leptin is secreted by adipocytes in response to feeding in order to suppress further food intake via an action at the hypothalamus. Thus, the critical locus of activity in this regulatory system occurs at the neuroendocrine level. This volume sets out to explore key aspects of this system. Following introductory chapters by Arthur Campfield, Joel Elmquist provides a survey of the anatomical basis for leptin activity in the hypothalamus. Some would argue that from an evolutionary point of view leptin is probably of greater relevance to the fasting than the fed state and this issue is explored in the chapter by Rexford Ahima. Chapters on interactions with NPY (Peter Widdowson and John Wilding) and POMC (Charles Mobbs and Tooru Mizuno) highlight the importance of pathways which subserve the adipostatic signal. The crucial role played by leptin in the regulation of reproduction is explained in the chapter by Clement Cheung, Donald Clifton and Robert Steiner. Finally, the intriguing possibility that the CNS itself may be a source of leptin is explored in the chapter by Michael Wilkinson. The important discoveries in this field that are highlighted in this volume are sure to be translated into advances in the management of obesity.

E. Ur, Halifax, Canada

Ur E (ed): Neuroendocrinology of Leptin.
Front Horm Res. Basel, Karger, 2000, vol 26, pp 1–11

Neurobiology of OB Protein (Leptin)

Introduction

L. Arthur Campfield

Department of Metabolic Diseases, Hoffmann-La Roche Inc., Nutley, N.J., USA

Introduction

The rapid elucidation of the properties, target tissues and actions of OB protein (also known as leptin), which is the product of the *ob* gene, has invigorated and energised obesity research as no other finding in this field has in the last 35 years. The circulating concentrations of OB protein are proportional to adiposity and increase with increasing levels of body fat [1]. The OB protein pathway is the long-sought hormonal signal pathway from adipose tissue to the brain that plays a critical role in the regulation of energy balance [2]. OB protein is a 16-kD polypeptide hormone that is secreted primarily from adipose tissue (and to a lesser extent bone marrow and placenta), circulates in the blood, bound to a family of binding proteins, enters the brain, binds to its receptor in hypothalamic nuclei and other brain areas and acts on central neural networks. Available evidence suggests that OB protein appears to play a major role in the control of body fat stores through co-ordinated regulation of feeding behavior, metabolism, neuroendocrine responses, autonomic nervous system and body energy balance in rodents, primates and humans [2].

The mechanisms in the brain responsible for determining the level at which body fat content is regulated in humans and other animals are not completely understood. A similar lack of knowledge exists for the mechanisms regulating the neuroendocrine rhythms supporting the adaptation to starvation and reproductive function in humans and other animals. Elucidation of the OB protein pathway within the brain has begun to provide important insights into these, and possibly other, mechanisms. If OB protein proves to be a useful tool to illuminate the mechanisms controlling body fat content and its corresponding decision rules or algorithms, it may provide the basis for a clearer understanding of the regulation of body fat content and energy balance [2]. When these mechanisms are understood at the molecular level, they should provide new targets for therapeutic interventions that will reduce and maintain body fat at reduced levels and, therefore, increase metabolic fitness, reduce risk factors and promote improved health of obese individuals. It is this hope that creates much of the excitement that greets each research advance in the understanding of the OB protein pathway [2].

In this chapter, the available knowledge about the OB protein pathway within the brain will be discussed. Studies conducted in laboratory animals and humans as well as in vitro experiments, have revealed that OB protein, by acting on diverse brain structures and mechanisms, regulates ingestive behavior, metabolism, neuroendocrine rhythms and controls body energy balance. The concept of reduced brain sensitivity to OB protein in obesity and the interaction of OB protein with other brain mechanisms will be summarised.

Obesity and the Neurobiology of OB Protein

Obesity is a major health problem throughout the world. Obesity is the most common nutritional disorder in the developed world and is associated with significant chronic metabolic diseases (hypertension, non-insulin-dependent diabetes mellitus (NIDDM), hypercholesterolemia) as well as stroke, sleep apnea, joint diseases and certain cancers. It is a complex, multifactorial disease characterised by behavioral, endocrine and metabolic alterations with an increasing prevalence [3]. Obesity is a cause of significant morbidity and is having an increasing negative impact on the health care systems in both the developed and developing world. Although treatment (e.g. diet, exercise, drugs) is available and most people can achieve medically significant weight loss (5–10% initial body weight), the long-term maintenance of that weight loss is, unfortunately, very rare. Thus, obesity remains a poorly managed medical condition that is a major cause of morbidity and mortality [3–5].

Ample evidence exists that obesity is, at its basis, a disease of biological dysregulation. The integration of multiple biological factors (including endocrine and metabolic factors), which are at least partially genetically determined, is thought to result in the steady-state body weight of an individual. When the steady-state weight is perturbed in either direction (increased, decreased), changes in body weight are resisted and corrected by robust physiological mechanisms in laboratory rodents and humans [6]. The physiological mechanisms that resist changes in body fat content (e.g. central neural networks, autonomic neural, metabolic, and neuroendocrine) are not completely known.

Results of available studies suggest that OB protein acts on the brain and probably plays a role in this 'resetting response' that is responsible for weight regain following weight loss [2].

OB Protein Biology in Humans

Numerous clinical studies of circulating OB protein concentrations have been performed. The basic facts can be summarised as follows: When obese subjects were compared to lean individuals, it was observed that the serum OB protein concentrations were higher in obese individuals, and OB protein concentrations increased with increasing percent body fat [1]. Thus, obese humans are not deficient in OB protein, but rather they have elevated circulating OB protein concentrations. Women have higher OB protein concentrations than men, even when corrected for the percent body fat. Subjects with NIDDM had lower OB protein levels than obese subjects but higher than lean subjects. When obese subjects lost weight by caloric restriction, OB protein concentration decreased and then rose slightly when the lower weight was maintained [1].

Measurements of the concentrations of OB protein in the cerebrospinal fluid (CSF) and the serum and the calculation of CSF/serum ratios of OB protein of lean and obese humans have been reported. The CSF concentrations of OB protein were correlated with BMI but were much lower than the serum concentrations. Although the serum concentrations of OB protein were much higher in obese individuals compared to lean subjects, the levels of OB protein in the CSF in obese humans were low and similar to those of lean subjects. This observation has lead to the suggestion that the brain uptake of OB protein and/or its appearance in the CSF compartment is defective in obese humans and it may be a component of the decreased sensitivity to OB protein [2, 7, 8].

Like much else in the OB protein pathway, the functional significance of low CSF concentrations of OB protein must also await additional experiments conducted in humans. However, the transport of radiolabelled OB protein into the brain of mice has been reported by Banks et al. [9]. In these studies, radiolabelled OB protein was injected intravenously into mice and they showed the presence of ^{125}I in the arcuate nucleus of the hypothalamus shortly after the injection. The rate of uptake of radiolabelled OB protein was decreased by coinjection of unlabelled OB protein, suggesting that the transport system for OB protein in mice may be saturable. A saturable transport system for OB protein in isolated human brain microvessels has been discovered and reported [10]. Many of the features of the brain transport system for OB protein appear to be similar to the transport system already described for brain insulin [2].

Human Mutations in the OB Protein Pathway

The global search for obese patients with mutations in the *ob* gene or the OB-R receptor has recently been successful. After screening thousands of obese individuals, mutations in the *ob* and *ob-r* genes have been identified in three different families. First, two cousins with severe obesity were identified with mutations in the *ob* gene and, as expected, they have little or no functional OB protein circulating in their blood. They were both normal weight at birth and then rapidly gained weight and became obese, just like the OB-protein-deficient obese mice [14]. Three members of another family have a homozygous missense mutation in the *ob* gene that results in low plasma OB protein concentration and severe obesity [15]. Three severely obese sisters in a large consanguineous family of Kabilian origin are homozygous for a splice-site mutation in the *ob-r* gene [16]. These discoveries strongly suggest that OB protein plays an important role in the regulation of body fat and body weight and reproductive function in humans.

OB Protein Receptor, OB-R

The OB protein receptor, OB-R, was rapidly identified and characterized following the identification of a central binding site for labelled OB protein in the choroid plexus and pia mater in *ob/ob, db/db* and lean mice as well as in lean and obese Zucker rats [11]. The OB-R has considerable homology with the GP130 subunit of the IL-6 receptor. It was concluded that OB-R has significant homology to a cytokine receptor [11]. OB-R was found to be expressed in the choroid plexus, the hypothalamus, many other brain areas as well as several peripheral tissues [2]. However, OB-R is a low abundance message and protein, making unambiguous detection of the receptor message and protein difficult. As a result of alternate splicing, OB-R exists in multiple forms. The two major forms are a short form, OB-R_S (with a truncated intracellular domain) and long form, OB-R_L (with the complete intracellular domain). The long form, OB-R_L, is thought to be the form that signals and mediates the biological effects of OB protein [11]. In situ hybridisation studies have demonstrated that the mRNA for OB-R_L is localised primarily to the hypothalamus (arcuate, lateral, ventromedial, dorsomedial nuclei) [12, 13].

Interaction of OB Protein with Other Brain Hormones and Neuropeptides

Research on the brain mechanisms involved in OB protein action has been stimulated by the long-lasting reductions in food intake and body weight

Table 1. Recent progress in neurobiology of OB protein

Neural activation by central and peripheral OB protein using *c-fos*-like activity in brains of mice and rats
Colocalisation studies of OB-R and NPY or POMC containing hypothalamic neurons
Clear evidence that OB protein modulates *npy, crh*, and *pomc* gene expression in hypothalamic neurons
Interactions between OB protein pathway and melanocortin signal pathways under active investigation
Demonstration that central and peripheral sensitivity to OB protein is reduced in DIO rats and mice
Localisation of OB-R_L in hypothalamic neurones; but majority in intracellular compartment
Further support that OB protein modulates 'hypothalamic-pituitary-target organ' axis
Clear evidence that OB protein plays a role in the control of reproduction
Successful treatment of two obese cousins with *ob* gene mutations with recombinant OB protein shows the operation of OB protein pathway in humans

of obese *ob/ob* mice observed following intracerebroventricular (ICV), OB protein administration [17, 18] together with the observation that the circulating OB protein concentration was proportional to body fat in humans [1]. As discussed in chapter 2, central administration of OB reduces food intake, body weight, alters metabolism and inhibits NPY-induced feeding of obese *ob/ob* mice, lean mice and rats [17, 18].

Some of the recent advances in the neurobiology of OB protein are listed in table 1. The neurotransmitters and neuropeptides that directly mediate the actions of OB protein have not yet been identified, but experimental evidence is consistent with OB acting through several brain mechanisms [neuropeptide Y (NPY), corticotropin-releasing hormone (CRH), pro-opiomelanocortin (POMC)] to co-ordinate the regulation of energy balance. Which, if any, neurotransmitters and neuropeptides are responsible for the decreased sensitivity to OB protein in obesity are not known. However, the neuronal network controlling energy balance and the descending sympathetic nervous system following peripheral and central administration of OB protein is beginning to emerge.

Patterns of activation of brain areas in response to central or peripheral administration of OB protein have been measured using *c-fos* immunoreactivity. Numerous studies have implicated hypothalamic nuclei [arcuate, ventromedial hypothalamus (VMH), dorsomedial hypothalamus (DMH), paraventricular hypothalamus (PVN)] and other brain areas thought to be involved in the control of energy balance [19, 20].

The first postulated major mediator of the actions of OB protein was NPY, a potent stimulator of feeding [21, 22]. This proposal was based on the inhibitory effects of OB protein on NPY gene expression and secretion [18] observed in early studies of biological activity of OB protein. We and others examined the interaction of brain administration of OB protein and NPY on the feeding behavior of *ob/ob* mice to test this hypothesis. We found that OB protein inhibited the expected feeding following NPY administration [23]. This indicates that OB protein can functionally antagonise the actions of exogenous NPY and suggests that the receptor-mediated actions of NPY on feeding are under the control of OB protein. These results are also consistent with other mediators, besides NPY, of OB protein action. This possibility has been strengthened by the report that normal weight mice lacking NPY respond to peripheral administration of OB protein [24].

When OB protein was administered into the third ventricle of the brain of lean rats, NPY gene expression in the arcuate nucleus over the subsequent 48 h was decreased, while CRH gene expression in the paraventricular nucleus and the expression of the POMC gene in the arcuate nucleus were increased [13]. However, other investigators have found that OB protein administration decreased CRH gene expression in the paraventricular nucleus [25]. Recent genetic and pharmacological research has provided strong evidence that the melanocortin 4 receptor (MC4-R) plays an important role in the regulation of body energy balance. In mice lacking the MC4-R, late-onset obesity and altered peripheral metabolism were observed [26]. This result, combined with experiments with MC4-R agonists (decreased food intake) and antagonists (increased food intake), indicates that the MC4-R lies in a new physiological pathway that normally inhibits food intake and fat storage, and plays an important role in late-onset obesity. It is widely believed that the endogenous ligand of the MC4-R is a product of POMC processing [27].

Colocalisation studies using selective antibodies have demonstrated that OB protein acts through OB-R on neurones that contain NPY, ACTH, CRH, POMC, somatostatin, galanin, tyrosine hydrolase, and MCH [28]. These findings support an integrative role for OB protein and/or the OB protein pathway.

A marked increase in the potency of OB protein was observed in adrenalectomized rats treated with OB protein [29]. These results provide support for a potentially important interaction between OB protein and the glucocorticoid system [2].

It has been demonstrated that OB protein could modulate synaptic transmission when applied to hypothalamic slice preparations from rat brain. This report suggested another mechanism to explain the behavioral and metabolic actions of OB protein [30]. Direct administration of OB protein on glucose-receptive ATP-sensitive potassium channels resulted in hyperpolarization of hypothalamic neurons [31]. Direct application of OB protein to VMH and

LH neurons resulting in both stimulation and inhibition of firing rates have been reported [20]. Future neurophysiological research will help evaluate if OB protein has a neuromodulatory action.

OB protein is not only linked to the regulation of energy balance but also may have a number of additional roles such as maintaining the normal neuroendocrine activity that is important in the adaptation to starvation and controlling stress responses and functions such as reproduction. These roles will be discussed in following chapters.

The OB Protein Pathway in the Brain

One possible role for OB protein is that of a modulator of gene expression and the resultant synthesis of one or more neurotransmitters and/or neuropeptides within the brain. Much of the research conducted on the OB protein pathway has been focussed on the identification of downstream genes that are regulated by OB protein. Three such genes are *npy*, *crh*, and *pomc* because OB protein has been shown to decrease the mRNA for NPY and increase mRNA for CRH and POMC mRNA in different areas of the hypothalamus [2].

Another possible role for OB protein is that of a modulator of synaptic transmission within the brain. Recent electrophysiological studies have supported this hypothesis. An alternative role for OB protein would be a 'co-ordinator' or 'organiser' of the seemingly disparate neurotransmitter and neuropeptide effects on, and responses to, ingestive behavior and body energy balance [2].

These three possible roles do not have to be, and are probably not mutually exclusive. Indeed, the 'co-ordinator' or 'organiser' function of OB protein within the brain may emerge from at least two distinct regulatory components mediated by the OB-R. One, the action of OB protein to regulate gene expression within areas of the brain critical for regulation of energy balance can be considered responsible for the 'long-term' or 'chronic' biological effects (e.g. critical molecules which determine the brain sensitivity to OB protein, 'settling point' for body fat content). Two, the action of OB protein to modulate synaptic transmission within the brain by altering the release or postsynaptic action of one or more neurotransmitters and/or neuropeptides can be considered responsible for the 'short-term', 'acute' or 'immediate' behavioral (suppression of food intake) and metabolic effects (decreased serum concentrations of insulin and glucose). This 'dual-function' mechanism of action, regulation of gene expression and modulation of ongoing cellular function such as neurosecretion, is a common and classical property of hormones [2].

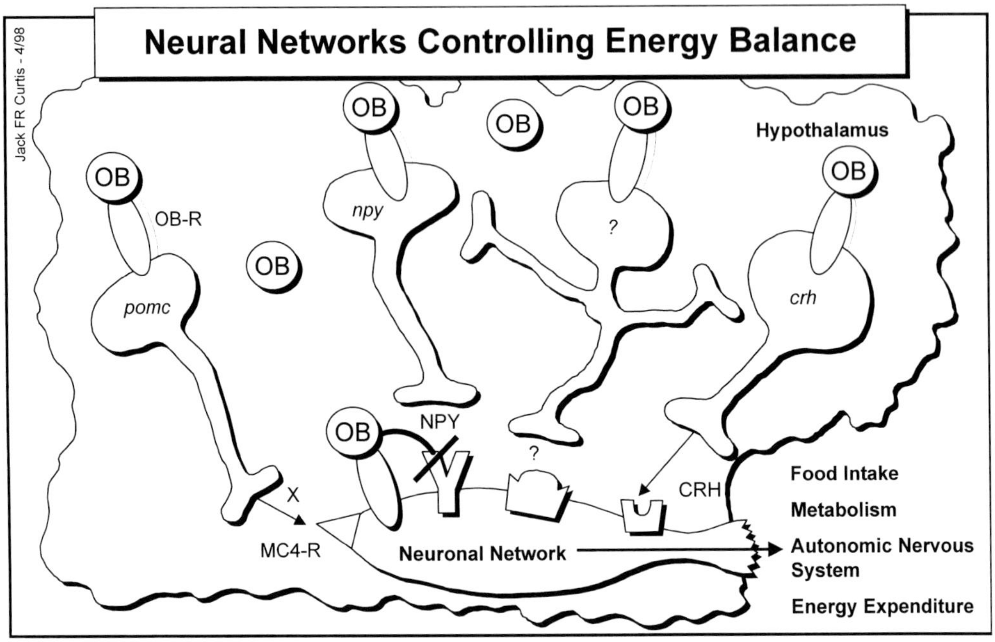

Fig. 1. Schematic diagram of hypothetical neural networks controlling energy balance. Several 'classes' of parallel neural pathways are represented by the four model neurones depicted. Each neurone is assumed to have the gene, synthesise and release one dominant type of neuropeptide – X, NPY, ?, CRH. Each neurone has the long form of the OB-R receptor (ellipses) and the 'final common pathway'. The neuronal network controlling ingestive behaviour, metabolism and energy balance, has distinct receptors for MC4-R, OB-R, NPY, ?, and CRH. OB protein is shown as circles containing 'OB'. Note that OB protein can express its biological action through four parallel paths, each mediated by a different neuropeptide – X, NPY, ?, CRH. The ability of OB protein to inhibit the actions of released NPY is also shown. Reprinted from Campfield and Smith [2].

One representation of this hypothesis at the level of the hypothalamic neurone is shown in figure 1. Several 'classes' of parallel neural pathways are represented by the four model neurones depicted. Each neurone shown is assumed to have the gene, synthesise and release one dominant type of neuropeptide – X, NPY, ?, CRH. Each neurone has the long form of the OB-R receptor (ellipses) and the 'final common pathway'. The neuronal network controlling ingestive behavior, metabolism and energy balance, has distinct receptors MC4-R, OB-R, NPY, ? and CRH. The demonstrated ability of OB protein to inhibit responses to NPY is shown by the heavy line blocking the NPY receptor. In this theoretical construct, OB protein would fulfil the role of 'conductor' integrating the distinct 'sections' of the 'orchestra' to behave as one while it plays the 'symphony'. In this analogy, the 'symphony' is the integrated behavioral, neuroendocrine and metabolic response of an individual, the 'orchestra' is the spatially distributed neural network controlling ingestive behavior, metabolism and energy balance, and the 'sections' would be the elements and subsystems of this neural network. This attractive hypothesis has been validated by several recent studies.

Concluding Remarks

The rapid characterisation of the OB protein signal pathway within the brains of laboratory rodents and humans has been significant for several reasons. OB protein has also demonstrated that the neurobiology and neuroendocrinology of ingestion, metabolism and energy balance have moved to the front of the research agenda. The unravelling of the brain mechanisms underlying the behavioral and metabolic actions of OB protein and responsible for the sensitivity of the brain to OB protein has become a major objective of neuroscience, neuroendocrinology and obesity research. The rapid evolution of the OB protein pathway has provided a series of important advances in the knowledge base of this field. This increasing knowledge base should help to pry open and shine a bright light into the very dark 'black box' that contains the mechanisms and decision rules or algorithms used in the brain for determining the level at which body fat content is regulated. When these mechanisms are understood at the molecular level, they will provide novel targets for discovery and development of new, safe, effective pharmacological treatment for obesity as adjuncts to diet and exercise in the future. It is hoped that these new therapeutic agents will reduce and maintain body fat at reduced levels and, therefore, increase metabolic fitness, reduce risk factors and promote improved health of obese individuals.

Acknowledgement

We thank Jack F. R. Curtis for assistance with the conceptualisation and execution of the artwork. We also thank our departmental colleagues and many collaborators for stimulating discussions and suggestions and Dr. Paul Burn for his interest and support.

References

1 Considine RV, Sinha MK, Heiman ML, Kriauciunas A, Stephens TW, Nyce MR, Ohannesian JP, Marco CC, McKee LJ, Baur TL, Caro JF: Serum immunoreactive-leptin concentrations in normal-weight and obese humans. N Engl J Med 1996;334:292–295.

2 Campfield LA, Smith FJ: Overview: Neurobiology of OB protein (leptin). Proc Nutr Soc 1998;57: 429–440.

3 Thomas PR (ed): Weighing the Options: Criteria for Evaluating Weight-Management Programs. Food and Nutrition Board, Institute of Medicine, Washington, National Academy Press, 1995.

4 Bjorntorp P, Brodoff BN (ed): Obesity. Philadelphia, Lippincott, 1992.

5 Campfield LA: Treatment options and the maintenance of weight loss; in Allison DB, Pi-Sunyer FX (eds): Obesity Treatment: Establishing Goals, Improving Outcomes, and Reviewing the Research Agenda. New York, Plenum Press, 1995, pp 93–95.

6 Leibe RL, Rosenbaum M, Hirsch J: Changes in energy expenditure resulting from altered body weight. N Engl J Med 1995;232:621–628.

7 Caro JF, Kolaczynski JW, Nyce MR, Ohannesian JP, Opentanova I, Goldman WH, Lynn RB, Zhang P-L, Sinha MK, Considine RV: Decreased cerebrospinal-fluid/serum leptin ratio in obesity: A possible mechanism for leptin resistance. Lancet 1996;348:159–161.

8 Schwartz MW, Peskind E, Raskind M, Boyko EJ, Porte D Jr: Cerebrospinal fluid leptin levels: Relationship to plasma levels and to adiposity in humans. Nat Med 1996;2:589–593.

9 Banks WA, Kastin AJ, Huang W, Jaspan JB, Maness LM: Leptin enters the brain by a saturable system independent of insulin. Peptides 1996;17:305–311.

10 Golden PL, Maccagnan TJ, Pardridge WM: Human blood-brain barrier leptin receptor: Binding and endocytosis in isolated human brain microvessels. J Clin Invest 1997:99:14–18.

11 Tartaglia LA, Dembski M, Weng X, Deng N, Culpepper J, Devos R, Richards JG, Campfield LA, Clark FT, Deeds J, Muir C, Sanker S, Moriarty A, Moore KJ, Smutko JS, Mays GG, Woolf EA, Monroe CA, Tepper RI: Identification and expression cloning of a leptin receptor (OB-R). Cell 1995;83:1263–1271.

12 Mercer JG, Hoggard N, Williams LM, Lawrence CB, Hannah LT, Trayhurn P: Localization of leptin receptor mRNA and the long form splice variant (Ob-Rb) in mouse hypothalamus and adjacent brain regions by in situ hybridization. FEBS Lett 1996;387:113–116.

13 Schwartz MW Seeley RJ, Campfield LA, Burn P, Baskin DG: Identification of targets of leptin action in rat hypothalamus. J Clin Invest 1996;98:1101–1106.

14 Montague CT, Farooqi IS, Whitehead JP, Soos MA, Rau H, Wareham NJ, Sewter CP, Diby JE, Mohammed SN, Hurst JA, Cheetham CH, Earley AR, Barnett AH, Prins B, O'Rahilly S: Congenital leptin deficiency is associated with severe early-onset obesity in humans. Nature 1997;387:903–908.

15 Strobel A, Issad T, Camoin L, Ozata M, Strosberg D: A leptin missense mutation associated with hypogonadism and morbid obesity. Nat Genetics 1998;18:213–215.

16 Clement K, Valisse C, Lahlou N, Cabrol S, Peeloux V, Cassuto D, Gourmelen M, Dina C, Chambaz J, Lacourte M, Basdevant A, Bougneres P, Lebouc Y, Froguel P, Guy-Grand B: A mutation in the human leptin receptor gene causes obesity and pituitary dysfunction. Nature 1998;392:398–401.

17 Campfield LA, Smith FJ, Guisez Y, Devos R, Burn P: Recombinant mouse OB protein: Evidence for a peripheral signal linking adiposity and central neural networks. Science 1995;269:546–549.

18 Stephens TW, Basinski M, Bristow PK, Bue-Valleskey JM, Burgett SG, Craft L, Hale J, Hoffmann J, Hsiung HM, Kriauciunas A, MacKeller W, Rosteck PR Jr, Schoner B, Smith D, Tinsley FC, Zhang X-Y, Heiman M: The role of neuropeptide Y in the antiobesity action of the obese gene product. Nature 1995;377:530–532.

19 Thiele TE, van Dijk G, Campfield LA, Smith FJ, Burn P, Woods SC, Bernstein IL, Seeley RJ: Central infusion of GLP-1, but not leptin, produces conditioned taste aversion in rats. Am J Physiol 1997;272:R726–R730.

20 Yokosuka M, Xu B, Pu S, Karla SP: The magnocellular part of the paraventricular hypothalamic nucleus (PVN): A selective site for inhibition of NPY-induced food intake by leptin. Soc Neurosci Abstr 1997;23:852.

21 Rohner-Jeanrenaud F, Cusin I, Sainsbury A, Zakrzewska K, Jeanrenaud B: Neuropeptide Y and leptin in lean and genetically obese fa/fa rats. Horm Metab Res 1996;28:642–648.

22 Ezzell C: Fat times for obesity research: Tons of new information, but how does it all fit together. J NIH Res 1995;7:39–45.

23 Smith FJ, Campfield LA, Moschera JA, Bailon P, Burn P: Feeding inhibition by neuropeptide Y. Nature 1996;382:307.

24 Erickson JC, Clegg KE, Palmiter RD: Sensitivity to leptin and susceptibility to seizures of mice lacking neuropeptide Y. Nature 1996;381:415–418.

25 Huang Q, Rivest R, Richard D: Effects of leptin on corticotropin-releasing factor (CRF) synthesis and CRF neuron activation in the paraventricular hypothalamic nucleus of obese (ob/ob) mice. Endocrinology 1998;139:1524–1532.

26 Huszar D, Lynch CA, Fairchild-Huntress V, Dunmore JH, Fang Q, Berkemeier LR, Gu W, Kesterson RA, Boston BA, Cone RD, Smith FJ, Campfield LA, Burn P, Lee F: Targeted disruption of the melanocortin-4 receptor results in obesity in mice. Cell 1997;88:131–141.

27 Fan W, Boston BA, Kesterson RA, Hruby VJ, Cone RD: Melanocortinergic inhibition of feeding behavior and disruption with an agouti-mimetic. Nature 1997;385:165–168.

28 Hakansson M-L, Brown H, Ghilardi N, Skoda RC, Meister B: Leptin receptor immunoreactivity in chemically defined target neurons of the hypothalamus. J Neurosci 1998;18:559–572.

29 Zakrzewska KE, Cusin I, Sainsbury A, Rohner-Jeanrenaud F, Jeanrenaud B: Glucocorticoids as counterregulatory hormones of leptin: Toward an understanding of leptin resistance. Diabetes 1997;46:717–719.

30 Glaum SR, Hara M, Bindokas V, Lee C, Polonsky K, Bell G, Miller R: Leptin, the obese gene product, rapidly modulates synaptic transmission in the hypothalamus. Mol Pharmacol 1996;50:230–235.

31 Spanswick D, Smith MA, Groppi VE, Logan SD, Ashford MLJ: Leptin inhibits hypothalamic neurons by activation of ATP-sensitive potassium channels. Nature 1997;390:521–525.

L. Arthur Campfield, PhD, Distinguished Research Leader, Department of Metabolic Diseases, Hoffmann-La Roche Inc., Nutley, NJ 07110 (USA)
Tel. +1 973 235 2787, Fax +1 973 235 8128, E-Mail L_Arthur.Campfield@Roche.com

Ur E (ed): Neuroendocrinology of Leptin.
Front Horm Res. Basel, Karger, 2000, vol 26, pp 12–20

Central Mechanisms Responsible for the Actions of OB Protein (Leptin) on Food Intake, Metabolism and Body Energy Storage

L. Arthur Campfield

Department of Metabolic Diseases, Hoffmann-La Roche Inc., Nutley, N.J., USA

The mechanisms in the brain responsible for determining the level at which body fat content is regulated in humans and other animals are not completely understood. A similar lack of knowledge exists for the mechanisms regulating the neuroendocrine rhythms supporting the adaptation to starvation and reproductive function in humans and other animals. Elucidation of the OB protein pathway within the brain has begun to provide important insights into these, and possibly other, mechanisms. If OB protein proves to be a useful tool to elucidate the mechanisms controlling body fat content and its corresponding decision rules or algorithms, it may provide the basis for a clearer understanding of the regulation of body fat content and energy balance [1, 2].

Since the cloning of the *ob* gene in December 1994 [3], research has progressed along five parallel paths: (1) regulation of *ob* gene expression in adipose tissue in mice, rats and humans; (2) characterization of the biological actions and definition of the elements of the OB protein pathway in lean and obese mice and rats; (3) studies of the biology of OB protein in lean and obese humans; (4) studies of the brain structures and mechanisms through which OB protein acts, and (5) effects of OB protein on reproduction. The key elements of the OB protein pathway are a transport system for OB protein to enter the brain, OB-R receptors in hypothalamic nuclei, and neural and neuroendocrine outputs to peripheral tissues as discussed in chapter 1 [1, 2].

In this chapter, we will discuss the available knowledge about the OB protein pathway. This understanding is based on in vitro experiments as well as studies conducted in laboratory animals and humans. This research has revealed that the OB protein, by acting on diverse brain structures and mechanisms, regulates ingestive behavior, metabolism, neuroendocrine rhythms and controls body energy balance. The role of OB protein in obesity and the concept of reduced brain sensitivity to OB protein will also be discussed.

Regulation of Energy Balance

Energy balance is the result of the control of ingestive behavior, energy expenditure and energy storage in adipose tissue [1]. In most adults, both body weight and body fat content remain constant over many years or decades despite a very large flux of energy intake and expenditure (approximately 1 million calories/year). Energy intake is a discrete process as individual meals are separated by intermeal intervals, while energy expenditure and storage in adipose tissue are continuous physiological processes. However, the complex molecular mechanisms by which discrete ingestive behavior, continuous energy expenditure and dynamic energy storage in adipose tissue are integrated and matched remain largely unknown. Two candidate signals are brain insulin and OB protein [1].

Control of Food Intake: A Behavioral Response Dependent on OB Protein

Feeding behavior is the result of the complex central nervous system (CNS) integration of central and peripheral neural, hormonal and neurochemical signals relating to brain and metabolic states. Meals are initiated, maintained and terminated by specific sets of these central and peripheral signals several times a day separated by intermeal intervals without food intake [1]. These signals include patterns of neural afferent traffic; metabolites (glucose); energy flux (fatty acid oxidation, ATP) and hormones (insulin concentrations in plasma and brain; OB protein concentrations in plasma and neuropeptide concentrations in brain). The brain structures and mechanisms involved in the detection of these signals and their mapping into altered feeding behavior are beginning to emerge. One hypothesis for this integrated neural, metabolic and hormonal control of food intake postulates the interaction of five classes of signals: (1) hypothalamic neuropeptides; (2) brain insulin; (3) brain OB protein; (4) metabolic signals including transient declines in blood glucose concentrations, and (5) ascending and descending neural inputs [1]. These signals interact and provide the central/peripheral integration necessary to regulate food intake and match energy intake to energy expenditure to maintain body energy balance and composition.

Many experimental studies in mice, rats and many other species demonstrate that several neuropeptides modulate food intake when injected centrally and, in some cases, peripherally. Among the neuropeptides that affect food intake are neuropeptide Y (NPY), galanin, cholecystokinin (CCK), corticotropin-releasing hormone (CRH), and enterostatin. Synaptic concentrations of central neuropeptides and classical monoamine neurotransmitters are thought to be modulated by the central representations of peripheral metabolic state and act on postsynaptic receptors to control energy intake. Although some investigators still seek the identity of the 'one' major neuropeptide controlling human feeding behavior, most of the field has adopted a 'parallel' model in which multiple neuropeptides are involved and each play a role in determining human feeding behavior [1].

Control of Energy Expenditure

Total daily energy expenditure can be partitioned into resting metabolic rate, dietary and cold-induced thermogenesis and the energy cost of physical activity. The energy cost of various kinds of voluntary physical activity (e.g. walking, jogging, swimming) and the thermogenesis due to digestion and absorption of food as well as thermogenesis induced by a cold environment have been calculated for both men and women as a function of lean body mass (or 'fat-free mass'). However, the regulation of resting metabolic rate is a complex function of energy intake, energy balance and hormonal and autonomic neural activity. When an individual reduces caloric intake and shifts into negative energy balance, the resting metabolic rate also decreases. This regulatory adaptation appropriately reduces obligatory energy expenditure when energy intake is reduced. However, this same adaptation causes a deceleration of weight loss following voluntary caloric restriction for the purpose of weight loss and contributes to the difficulty of maintaining weight loss, once achieved, over time [1].

Experimental Models of Obesity

Altered OB Protein Pathway: Obese ob/ob *and* db/db *Mice*

The obese *ob/ob* mice were discovered on the C57BL/6J background in the 1950s at the Jackson Laboratories (Bar Harbor, Me., USA). This mutation results in profound obesity. The *db/db* mouse, which arose on the C57BL/KsJ background, is similarly obese and is also characterized by hyperglycemia. When the *ob* gene was transferred to the C57BL/KsJ background, an almost identical phenotype to *db/db* mice was observed. Likewise, when the *db* gene was transferred to the C57BL/6J background, an obese phenotype almost identical to *ob/ob* mice resulted [4]. Since the cloning of the *ob* and *ob-r* genes, it has become clear that the *ob/ob* mouse has a mutated *ob* gene and produces no OB protein from its fat cells; it is a 'natural' knockout and is extremely sensitive to OB protein. The *db/db* mouse has a mutation in the *ob-r* gene that causes all its OB-R to be the 'short' form; thus, it is totally unresponsive to OB protein [1, 2].

Diet-Induced Obesity

A model with potentially more applicability to human obesity is diet-induced obesity (DIO) in mice and rats. When very palatable, high-fat diets (remarkably similar to chocolate chip cookie dough without the chocolate chips – mainly composed of sugar and shortening added to powdered rodent food) are given to normal mice and rats, diet-induced obesity occurs in approximately 60–75% of some, but not all, strains of mice and rats. Animals will rapidly gain weight over a 4- to 6-week period, and have increased circulating concentrations of insulin and OB protein, and become very obese when fed these diets [5]. These diet-induced obese animals will also have decreased sensitivity to the actions of circulating insulin and OB protein. The operational definition for diet-induced obesity is a body weight > average body weight of animals fed normal food +2 SD. Diets composed of high fat content, high energy density, or multiple palatable items that are presented simultaneously (called 'supermarket' or 'cafeteria' diets) can produce diet-induced obesity in otherwise normal rodents. Strains that 'resist' diet-induced obesity and its metabolic alterations (e.g. SHR mice) and individual 'resistant' members of a susceptible strain are being studied because they may provide important biological clues to help reduce weight gain, and regain, in susceptible humans. Diet-induced obesity is an experimental model that is very relevant for studying obesity, particularly with regard to the OB protein pathway. This is because it offers us a chance to study a polygenic obese state in animals that may mimic the situation in most obese human subjects – the interaction of genetic predisposition, rather than mutation, with the environment.

Biological Activity of OB Protein in Mice and Rats

Study of the biology of OB protein began when biologically active forms of recombinant mouse and human OB protein were administered to obese *ob/ob* mice. Food intake and body weight of obese *ob/ob* mice were reduced following administration of OB protein peripherally (i.p., i.v.) or centrally (ICV) [1, 2, 6]. These are the results that we reported in *Science* in July 1995, together with a group from Amgen [7] and Rockefeller [8] that attracted both scientific and lay media attention.

Sustained, dose-related reductions in food intake and body weight of *ob/ob* mice compared to mice treated with vehicle control solution were observed following the repeated i.p. administration of OB protein (2 injections separated by 3 h of 1.5 or 3 μg/mouse/day) for 5 days.

Recombinant mouse OB protein did not reduce daily food intake or body weight of obese *ob/db* mice when similar repeated i.p. (6 μg/mouse) dose studies were conducted. Similar results were reported when recombinant mouse OB protein was injected i.p. in *ob/ob* and *db/db* mice for up to 28 days [74, 75] and later by several groups [1, 2]. These studies demonstrated that repeated i.p. administration of OB protein reduces daily food intake and body weight in obese *ob/ob* and DIO, but not in *db/db* mice. Reductions in plasma glucose, insulin and lipids were also observed following administration of OB protein [9].

Cumulative 7-hour food intake was reduced in a dose-related manner (0–45%) following the administration of single i.v. doses of recombinant mouse OB protein (0.1–3 μg/mouse) to overnight fasted *ob/ob* mice. Recombinant mouse OB protein was then administered i.v. to chronically cannulated lean rats. When a single dose (30 μg/rat or 0.12 mg/kg) of mouse OB protein was injected i.v. into overnight fasted lean rats, cumulative 7-hour food intake was reduced by 45% and body weight gain during the 24-h postinjection period was reduced.

To determine if the brain was a target of OB protein, recombinant mouse OB protein was injected (ICV) (lateral ventricle) to *ob/ob* and lean (+/?) mice in our first study of the biological activity of OB protein [6]. These mice were implanted with chronic lateral ventricle cannulas. Single ICV injections of 0.001–1 μg/mouse in 1 μl were administered to overnight-fasted *ob/ob* and lean (+/?) mice. Cumulative 7- and 24-hour food intake was suppressed in a dose-related manner. Postinjection body weight gains were also reduced in a dose-related manner compared to mice injected with vehicle control. These results have been confirmed by several investigators [2].

However, when the effective dose in lean rats of OB protein (3.5 μg) was placed in the third ventricle in obese Zucker rats, no behavioral effects were observed [10]. In contrast, others [11, 12] reported that higher doses of OB protein did suppress food intake of obese Zucker rats. Further studies will be needed to determine the impact of the point mutation in the OB-R of the obese Zucker rats on its sensitivity to OB protein.

When a compound or protein administered peripherally or centrally results in reduced food intake and the loss of body weight, the possibility that these behavioral effects are due to nonspecific action or illness produced by the test substance must be considered. This possibility is assessed by conditioned taste aversion studies [13]. In studies with OB protein, lean rats were offered a novel taste paired with the ICV injection of OB protein, the i.p. injection of LiCl or no treatment. This decreased preference for a taste paired with a toxic substance is called a conditioned taste aversion. In contrast, ICV treatment with OB protein had no effect on the consumption of saccharin or water. These studies demonstrated that OB protein does not support the development of a conditioned taste aversion. These results provide strong support that the reduction in food intake and body weight observed following the administration of OB protein are specific biological effects of the hormone.

The results of these and many other experiments, taken together, provide further support for the hypothesis that a circulating protein-based signal, secreted from adipose tissue, acts on central neuronal networks and suggests that OB protein plays an important role in the regulation of ingestive behavior and energy balance (see chapter 1) [1, 2]. The duration of action of OB protein appears longer than those of other neuropeptides that modulate ingestive behavior (CCK, NPY, galanin) and is similar to that of centrally administered insulin. The behavioral and physiological effects following central administration of OB protein into the lateral ventricle suggest that OB protein can act directly on the neural networks in the brain that regulate ingestive behavior and energy balance.

Effects of OB Protein on Metabolic Rate

A very well-established adaptation to caloric restriction is a fall in metabolic rate. This reduction in metabolic rate has been repeatedly observed in rodents, larger mammals, primates and humans (14–16]. This adaptation often frustrates voluntary attempts at weight loss in humans.

It has been shown that following central administration of OB protein to animals undergoing caloric restriction, the expected fall in metabolic rate in response to caloric restriction was attenuated [17, 18]. These findings suggest that one of the biological actions of OB protein is to regulate adjustments in metabolic rate. The central and peripheral mechanisms and their integration underlying this regulation of metabolic rate are at present unknown. If this biological effect of OB protein translates to humans, then therapeutic activation of the OB protein pathway could revolutionize the treatment of obesity just by keeping metabolic rate elevated during caloric restriction.

Role of OB Protein in Obesity and the Concept of Reduced Sensitivity to OB Protein in Obesity

Based on the available results in animal models and humans, most human obesity is probably not due to a deficiency of OB protein, but instead is associated with, and may be due to, central and/or peripheral resistance or decreased sensitivity to OB protein. Strong support for the concept of reduced sensitivity to OB protein in obesity is provided by the observation of elevated OB protein concentrations in the blood of obese individuals and the experimental result that higher doses are required to effect feeding behavior, metabolism and body fat in DIO mice [6, 19]. At one extreme of the continuum of OB protein responsiveness is the obese *db/db* mouse, with elevated OB protein levels, which is totally unresponsive to OB protein, while at the other end is the obese *ob/ob* mouse, OB protein-deficient, which is very responsive to OB protein.

Movement along the continuum of OB protein sensitivity as a function of adiposity and back again has been demonstrated in studies of DIO mice. When lean AKR/J mice are fed a high-fat, energy-dense diet, they become obese with elevated OB protein and insulin concentrations [2, 6]. Our demonstration that these DIO obese mice required higher IP doses of OB protein to reduce food intake contributed to the concept of OB protein resistance. Recently, we have shown that DIO mice have decreased central sensitivity to OB protein [20]. In these studies, we determined the sensitivity to ICV injection of mouse OB protein in DIO mice after changes in diet. These results demonstrate that expansion of the adipose tissue mass, as a result of high-fat-diet feeding, is associated with decreased brain sensitivity to OB protein injected ICV. Thus, these studies with DIO mice indicate that the brain sensitivity to OB protein of the neural network in the brain controlling energy balance is decreased by weight gain and can be reversed by weight loss.

OB protein is not only linked to the regulation of energy balance but also may have a number of additional roles such as maintaining the normal neuroendocrine activity that is important in the adaptation to starvation and controlling stress responses and functions such as reproduction. Starved mice and OB protein-deficient *ob/ob* mice were found to have similar neuroendocrine abnormalities including an activated hypothalamic-pituitary-adrenal response and depressed thyroid function. In addition, male mice had low levels of testosterone and delayed ovulation was observed in female mice. When fasted mice were treated with OB protein, these neuroendocrine abnormalities were reversed. Administration of OB protein to female *ob/ob* mice restored reproductive function to near-normal and the mice became pregnant and successfully carried litters to term [21].

Significance of OB Protein

The rapid characterization of the OB protein signal pathway within the brains of laboratory rodents and humans has been significant for several reasons. First, it has provided additional evidence that obesity is a disease with a biological basis, including major endocrine and metabolic components, and it has stimulated hormonal and metabolically oriented obesity research. Second, it has demonstrated that molecular biology has reached the study of obesity, ingestive behavior, metabolism and has changed these fields forever. Molecular methods and techniques have now become part of the required tools in obesity research. Third, OB protein has also demonstrated that the neurobiology and neuroendocrinology of ingestion, metabolism and energy balance have been moved to the front of the research agenda.

Each advance in our understanding of the OB protein brings us closer to the development of an effective and safe drug that will increase the sensitivity of the OB pathway in brain of obese individuals to their own more-than-adequate levels of OB protein. Such an increase in the sensitivity of the OB protein pathway to an individual's own OB protein should assist them to reduce and then chronically maintain a lower amount of body fat that will result in improved health.

Acknowledgment

We thank our departmental colleagues and many collaborators for stimulating discussions and suggestions.

References

1 Campfield LA, Smith FJ, Burn P: OB protein: A hormonal controller of central neural networks mediating behavioral, metabolic and neuroendocrine responses. Endocrinol Metab 1997;4:81–102.

2 Campfield LA, Smith FJ: Overview: Neurobiology of OB protein (leptin). Proc Nutr Soc 1998;57: 429–440.

3 Zhang Y, Proenca R, Maffei M, Barone M, Leopold L, Friedman JM: Positional cloning of the mouse *obese* gene and its human homologue. Nature 1994;372:425–431.

4 Coleman DL: Obese and diabetes: Two mutant genes causing diabetes-obesity syndromes in mice. Diabetologia 1978;14:141–148.

5 West DB, York B, Waguespack J, Goudey-Lefevre J, Price AR: Genetics of dietary obesity in AKR/J × SWR/J mice: Segregation of the trait and identification of a linked locus on chromosome 4. Mammal Genome 1994;5:546–552.

6 Campfield LA, Smith FJ, Guisez Y, Devos R, Burn P: Recombinant mouse OB protein: Evidence for a peripheral signal linking adiposity and central neural networks. Science 1995;269:546–549.

7 Pelleymounter M, Cullen M, Baker M, Hecht R, Winters D, Bone T, Collins F: Effects of the obese gene product on body weight regulation in *ob/ob* mice. Science 1995;269:540–543.

8 Halaas JL, Gajiwala KS, Maffei M, Cohen SL, Chait BT, Rabinowitz D, Lallone RL, Burley SK, Friedman JM: Weight-reducing effects of the plasma protein encoded by the obese gene. Science 1995;269:543–546.

9 Schwarz MW, Baskin DG, Bukowski TR, Kuijper JL, Foster D, Lasser G, Prunkard DE, Porte D Jr, Woods SC, Seeley RJ, Weigle DS: Specificity of leptin action on elevated blood glucose levels and hypothalamic neuropeptide Y gene expression in *ob/ob* mice. Diabetes 1996;45:531–535.

10 Seeley RJ, van Dijk G, Campfield LA, Smith FJ, Burn P, Nelligan JA, Bell MS, Baskin DG, Woods SC, Schwartz MW: Intraventricular leptin reduces food intake and body weight of lean rats but not obese Zucker rats. Horm Metabol Res 1996;28:664–668.

11 Rohner-Jeanrenaud F, Cusin I, Sainsbury A, Zakrzewska K, Jeanrenaud B: Neuropeptide Y and leptin in lean and genetically obese fa/fa rats. Horm Metabol Res 1996;28:642–648.

12 Lin L, Truett GE, Levans N, York DA: Central, but not peripheral administration of leptin reduced the food intake and body weight in Zucker fatty and lean rats. Obes Res 1996;4:1S.

13 Thiele TE, van Dijk G, Campfield LA, Smith FJ, Burn P, Woods SC, Bernstein IL, Seeley RJ: Central infusion of GLP-1, but not leptin, produces conditioned taste aversion in rats. Am J Physiol 1997;272:R726–R730.

14 Bjorntorp P, Brodoff BN (eds): Obesity. Piladelphia, Lippincott, 1992.

15 Campfield LA: Treatment options and the maintenance of weight/loss; in Allison DB, Pi-Sunyer FX (eds): Obesity Treatment: Establishing Goals, Improving Outcomes, and Reviewing the Research Agenda. New York, Plenum Press, 1995, pp 93–95.

16 Le Magnen J: The metabolic basis of the dual periodicity of feeding in rats. Behav Brain Sci 1980; 4:561–607.

17 Speakmar JR, Trayhurn P, Rayner DV: Leptin inhibits the starvation responses of mice. Int J Obesity Relat Met Dis 1998;22:S177.

18 Doring H, Schwarzer K, Nuesslein-Hildesheim B, Schmidt I: Leptin selectively increases energy expenditure of food-restricted lean mice. Int J Obes Relat Metab Disord 1998;22:83–88.

19 Considine RV, Sinha MK, Heiman ML, Kriauciunas A, Stephens TW, Nyce MR, Ohannesian JP, Marco CC, McKee LJ, Baur TL, Caro JF: Serum immunoreactive-leptin concentrations in normal-weight and obese humans. N Engl J Med 1996;334:292–295.

20 Campfield LA, Smith. FJ, Yu J, Renzetti M, Simko B, Baralt M, Mackie G, Tenenbaum R, Smith W: Dietary obesity induces decreased central sensitivity to exogenous OB protein (leptin) which is reversed by weight loss. Soc Neurosci Abstr 1997;23:815.

21 Chehab F, Lim M, Lu R: Correction of the sterility defect in homozygous obese female by treatment with human recombinant leptin. Nat Genet 1996;12:318–320.

L. Arthur Campfield, PhD, Distinguished Research Leader, Department of Metabolic Diseases, Hoffmann-La Roche Inc., Nutley, NJ 07110 (USA)
Tel. +1 973 235 2787, Fax +1 973 235 8128, E-Mail L_Arthur.Campfield@Roche.com

Ur E (ed): Neuroendocrinology of Leptin.
Front Horm Res. Basel, Karger, 2000, vol 26, pp 21–41

Anatomic Basis of Leptin Action in the Hypothalamus

Joel K. Elmquist

Department of Neurology, Beth Israel Deaconess Medical Center, and
Program in Neuroscience, and Department of Medicine and
Division of Endocrinology, Beth Israel Deaconess Medical Center,
Harvard Medical School, Boston, Mass., USA

In the past few years the molecular basis for several obesity syndromes have been discovered. Prominent among them are the hormone leptin and its receptor. These observations have rapidly and dramatically increased our understanding of the pathophysiology of obesity and related disorders. Leptin, the product of the *ob* gene [1], is produced by white adipose tissue and affects feeding behavior, thermogenesis, and neuroendocrine status. Total absence of leptin (*ob/ob* mice) or lack of response to leptin (*db/db* mice) causes morbid obesity, diabetes, hypogonadism, hypothermia, and hypercorticosteronemia [2]. Replacement with exogenous leptin normalizes these abnormalities [3–5].

The profound and varied effects of leptin on multiple systems illustrates the importance of understanding the basic mechanisms of leptin interactions with the central nervous system. The behavioral, neuroendocrine, and autonomic effects of leptin are mediated by the hypothalamus and recent studies have begun to identify some of the complex circuitry involved in leptin biology. This review will address the effects of leptin on multiple neuroendocrine systems and will also discuss some of the identified neuroanatomic pathways and the neurotransmitter systems engaged by circulating leptin.

Leptin Regulates Neuroendocrine Systems

Total lack of leptin or leptin signalling in rodents and humans causes morbid obesity that is accompanied by a constellation of neuroendocrine abnormalities. Starvation, a time of low energy stores, leads to a fall in serum leptin levels, and has profound effects on several neuroendocrine systems including activation of the hypothalamo-pituitary-adrenal (HPA) axis, inhibition of the growth hormone and thyroid axes, and inhibition of reproductive function [2, 6–8]. Therefore, lack of leptin has many physiological responses that are also found in a state of starvation. Interestingly, many of these starvation-induced endocrine and autonomic changes are blocked or blunted by pretreatment with systemic leptin [6]. The dose needed to reverse these abnormalities is lower than needed to induce weight loss in normal rodents. These observations have led to the suggestion that circulating leptin may have evolved to signal the brain that energy stores are sufficient and that a lack of leptin may be responsible for multiple neuroendocrine abnormalities caused by starvation (see chapter 4).

Leptin and the HPA Axis

Elevated levels of corticosterone and ACTH are characteristic of *ob/ob* mice. Chronic administration of leptin to *ob/ob* mice decreases the plasma corticosterone levels [5, 9]. As mentioned, starvation stress in normal rodents also activates the HPA axis. Leptin administration during the fast, markedly blunts the rise in corticosterone and ACTH [6]. Leptin also appears to be able to inhibit acute activation of the HPA axis in another stress model. Specifically, restraint stress typically activates the HPA axis with an elevation of both ACTH and corticosterone. Leptin administration prior to the restraint blunts the rise in both ACTH and corticosterone [10]. To elucidate possible mechanisms for leptin's regulation of the HPA axis, Heiman et al. [10] investigated whether leptin may affect the release of corticotropin-releasing hormone (CRH) from the hypothalamus. They demonstrated that leptin inhibits hypoglycemia-induced secretion of CRH release from perfused hypothalamic slices.

Other studies also suggest that leptin may affect the function of hypothalamic CRH systems in vivo. Huang and colleagues found that leptin has inhibitory effects on the HPA axis as well. Specifically, leptin administration decreased CRH mRNA and expression of Fos protein in the paraventricular nucleus of the hypothalamus, and lowered serum corticosterone of *ob/ob* mice [11]. However, not all data suggests that leptin inhibits CRH expression. In fact, Schwartz et al. [12] found that intracerebroventricular (ICV) leptin administration to rats during a fast increased CRH mRNA in the paraventricular nucleus. Currently, it is unclear whether this increase in CRH is accompanied by an activation of the HPA axis. However, it should be noted that the paraventricular hypothalamic nucleus is made up of several populations of neurons that do not project to the median eminence to regulate ACTH secretion but are capable of making CRH [13].

Leptin and the Growth Hormone and Thyroid Axes

Growth hormone (GH) levels are influenced by body weight and decreased by starvation. Recent experimental evidence suggests that leptin regulates GH levels as well. Specifically, inhibition of leptin action by central administration of a leptin-neutralizing antisera decreased spontaneous GH secretion [14]. Leptin administration to fetal rat neurons in culture led to a time dependent decrease in basal somatostatin secretion and somatostatin mRNA levels. Finally, leptin also inhibited low-glucose-induced somatostatin secretion in perfused adult hypothalamic slices [15]. These data indicate that leptin can influence the growth hormone axis by regulating hypothalamic somatostatin gene expression.

Evidence is also accumulating that leptin regulates the thyroid axis. Most data to date exists using fasting paradigms. Fasting induces a number of changes in thyroid status that includes low triiodothyronine (T3) and thyroxine (T4) levels and a decreased level of thyroid-stimulating hormone (TSH). This response is thought to involve a decrease in thyrotropin-releasing hormone (TRH) mRNA levels in the paraventricular nucleus of the hypothalamus [16]. Ahima et al. [6] demonstrated that leptin administration prevents the fasting-induced suppression of the thyroid axis. Subsequently, Legradi et al. [17] found that systemic administration of leptin blocked the fasting-induced fall of proTRH mRNA in rats. Therefore, leptin appears to regulate the levels of proTRH mRNA in the paraventricular nucleus. The mechanisms for this regulation remain to be determined.

Leptin and the Reproductive Axis

An important connection between nutrition and reproduction has long been noted. Emerging evidence indicates that leptin may be an important link between nutritional state and reproductive capacity. The first observation supportive of this hypothesis is that *ob/ob* mice fail to undergo puberty, a process that is restored by administering leptin [18, 19]. In normal mice, leptin administration from the time of weaning accelerates the onset of puberty [20–22]. Additionally, in boys, leptin levels appear to peak at or before the time of puberty onset when studied in a longitudinal fashion through the peripubertal years [23]. However, in monkeys there is no increase in leptin levels prior to puberty [24]. The mechanisms for this regulation are not clear but clues are beginning to emerge. Yu and colleagues investigated the effects of leptin on hypothalamic and pituitary release of GnRH, LH and FSH. They found that leptin stimulated GnRH release from hypothalamic explants and FSH and LH release from anterior pituitaries of adult male rats in vitro and released LH, but not FSH, in vivo [25]. Their results indicate that leptin plays an important role in controlling gonadotropin secretion by stimulating the hypothalamus and pituitary. Taken together, these findings are consistent with the possibility that leptin is a metabolic signal that acts on the gonadotropin-releasing hormone system, either directly or indirectly [18–22, 26].

Distributions of Leptin Receptors

The aforementioned effects of leptin on multiple neuroendocrine systems demonstrates that leptin mediates its effects via the hypothalamus; however, the neuroanatomic basis for this remain unclear. Several splice variants of the leptin receptor (OB-R) have been identified, some of which localize to the brain. The predominant form is the ‘short form’ and is widely expressed in multiple tissues including the brain [27]. Another splice variant encodes a protein with a longer cytoplasmic domain (OB-Rb) and is often referred to as the ‘long form’ of the receptor. This form is highly expressed in the hypothalamus [12, 28–33]. Interestingly, mutations of this protein result in the obese phenotype of the *db/db* mouse and the Zucker rat indicating that leptin signaling via this receptor in the hypothalamus is critical in energy homeostasis [34–36].

Anatomic studies indicate that OB-Rbs localize to several hypothalamic nuclei with the highest in nuclear groups of the ventrobasal hypothalamus [12, 29, 31]. We systematically examined distributions of mRNA of leptin receptor isoforms in the rat brain by using a probe specific for the long form and a probe recognizing all known forms of the leptin receptor [32]. The mRNA for the long form of the receptor (OB-Rb) localized to selected nuclear groups in the rat brain. Within the hypothalamus, dense hybridization was observed in the arcuate, dorsomedial, ventromedial, and ventral premamillary nuclei (fig. 1, 2). Within the dorsomedial nucleus (DMH), particularly intense hybridization was observed in the caudal regions of the nucleus ventral to the compact formation (fig. 2F). Receptors were preferentially localized to the dorsomedial division of the ventromedial nucleus (fig. 1A). Hybridization accumulated throughout the arcuate nucleus extending from the retrochiasmatic region to the posterior periventricular region. Moderate hybridization was also observed in the periventricular hypothalamic nucleus, lateral hypothalamic area, medial mammillary nucleus, and posterior hypothalamic nucleus.

Interestingly, we found extrahypothalamic distributions of leptin receptors in the rat brain as well. These regions included the nucleus of the lateral olfactory tract, the substantia nigra pars compacta and several thalamic nuclei. The thalamic groups included the mediodorsal, ventral anterior, ventral

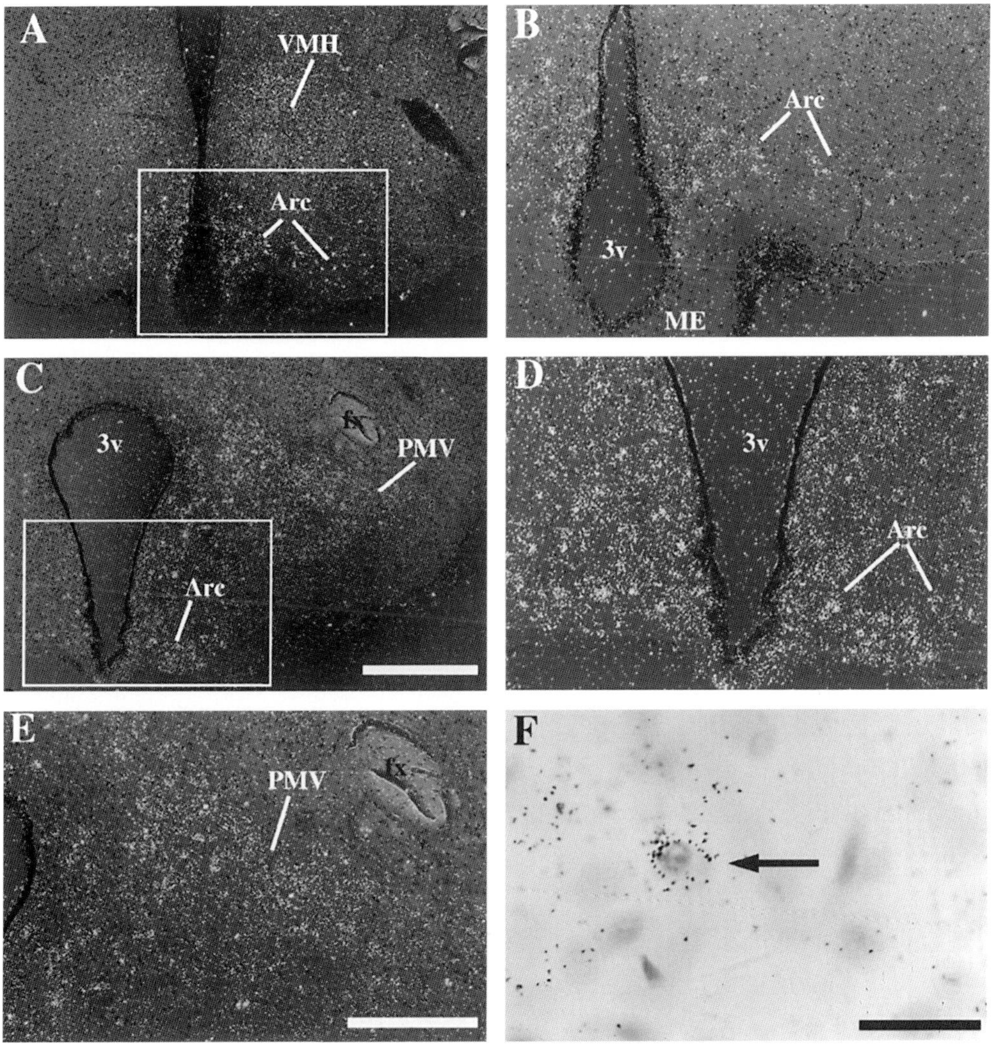

Fig. 1. A series of photomicrographs demonstrate the expression of leptin receptor mRNA in the rat hypothalamus. Hybridization of mRNA for the long form of the leptin receptor (OB-Rb) localizes to the arcuate hypothalamic nucleus (Arc; *A–D*, *F*), ventromedial hypothalamic nucleus (VMH; *A*), and the ventral premammillary nucleus (PMV; *C*, *E*). An arrow demonstrates specific hybridization over a presumed neuron in the Arc in a thionin counterstained section (*F*). *A* and *C* are at same magnification (bar = 500 μm); *B*, *D*, and *E* are at same magnification (bar = 300 μm); *F* bar = 25 μg. Reprinted with permission from Elmquist et al. [32].

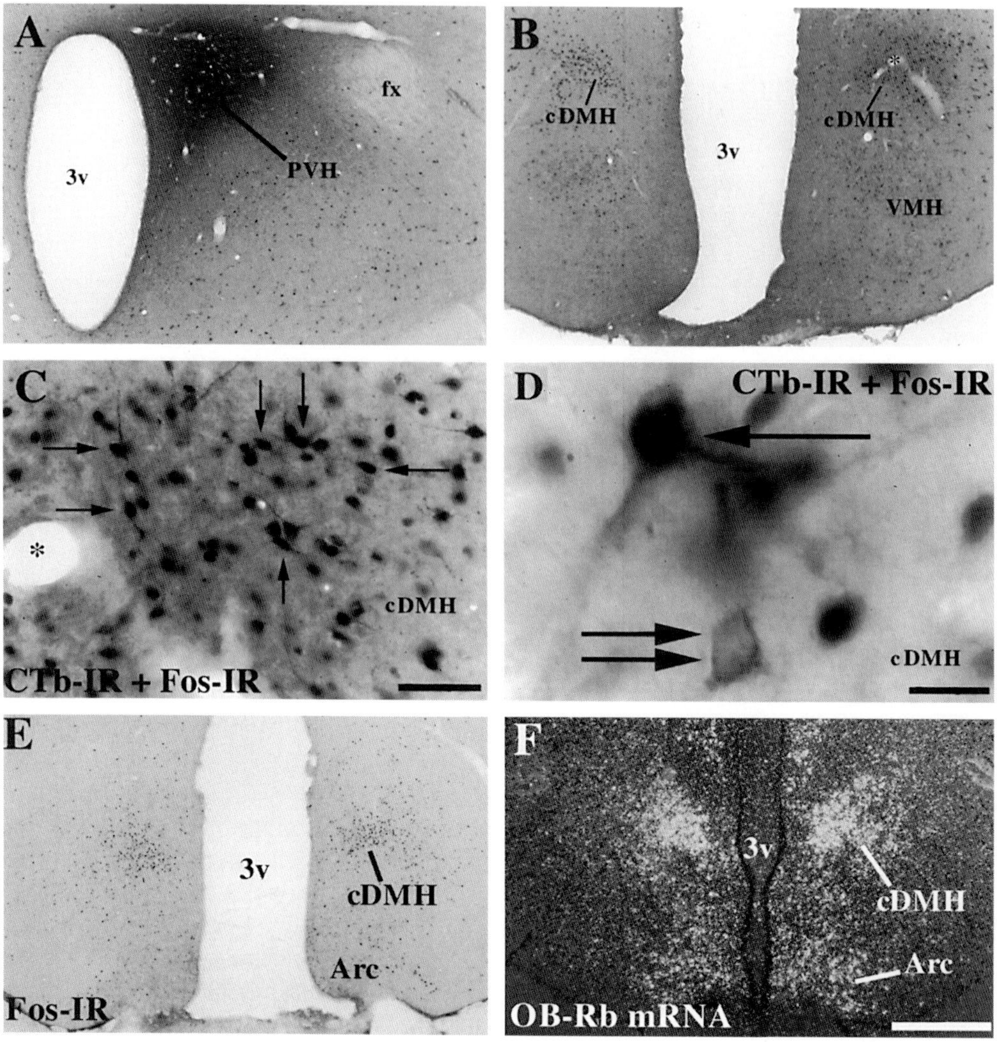

Fig. 2. Photomicrographs demonstrating the distribution of Fos-like immunoreactivity (Fos-IR; black nuclei), and CTb-like immunoreactivity (CTb-IR; retrogradely labeled cells; cytoplasmic staining), 2 h following intravenous leptin (1.0 mg/kg) and injections of CTb into paraventricular hypothalamic nucleus (PVH; *A*). Following injections of CTb into the PVH, many double-labeled cells are observed in the caudal dorsomedial hypothalamic nucleus (cDMH; *B–D*, use asterisk for orientation). Double-labeled cells are denoted by arrows, and single-labeled CTb cells by double arrows. Note the patterns of hybridization of mRNA for the long form of the leptin receptor (OB-Rb) and that the leptin induced Fos-IR localizes in the cDMH. *A*, *B*, *E*, *F* are at the same magnification, bar = 1.0 mm; *C* bar = 50 μm; *D* bar = 10 μm. Arc = Arcuate nucleus of the hypothalamus; fx = fornix; 3v = third ventricle. Reprinted with permission from Elmquist et al. [83].

medial, submedial, ventral posterior, and lateral dorsal thalamic nuclei. Hybridization was also observed in the medial and lateral geniculate nuclei. Finally, intense hybridization was observed in the Purkinje and granular cell layers of the cerebellum. Extrahypothalamic leptin receptors have also been described in the rodent and human brain by other investigators [32, 37–41]. These receptors are found in regions including the cerebellum, thalamus, parabrachial nucleus, and nucleus of the solitary tract. Currently, the role of extrahypothalamic leptin receptors is unknown. One possibility is that an undiscovered ligand for these receptors exists within the brain. Alternatively, leptin may act at these receptors to regulate diverse sensory and motor functions unrelated to body weight regulation. The role of these receptors remains to be determined but may provide interesting insights into the role of leptin in the maintenance of homeostasis.

A probe recognizing all known forms of the leptin receptor hybridized to all of the sites within the brain that were labeled with the long form specific probe. Additionally, using the all forms probe intense hybridization was observed in the choroid plexus, meninges, and also surrounding blood vessels. The presence of leptin receptor mRNA in the meninges and the microcirculation raises the intriguing possibility that leptin receptors at one or all of these sites are responsible for transporting leptin into or out of the CNS. Currently, it is not clear how circulating leptin gains access to the cell groups we identified. This is of great interest as white adipose tissue appears to be the major source of leptin in non-pregnant females [1, 42]. Therefore, mechanisms must exist to transport leptin across the blood-brain barrier to allow leptin access to neuronal populations within the CNS containing long-form leptin receptors.

Similar to many circulating hormones, leptin may signal the brain by interacting with circumventricular organs. Circumventricular organs are structures that line the ventricular system and lack a blood-brain barrier. They include the subfornical organ, organum vasculosum of the lamina terminalis, median eminence, and the area postrema [43]. It is plausible that circulating leptin may enter the brain through the median eminence to directly bind to receptors in the ventral basal hypothalamus including the overlying arcuate, ventromedial, and dorsomedial nuclei. However, transport via the circumventricular organs does not account for leptin reaching structures such as the cerebellum and thalamus. Rather, the potential for transport of leptin by short forms of the receptor located in cells lining cerebral blood vessels and in cells of the choroid plexus may underlie the ability of leptin to reach deeper brain structures. Indeed, transport of radiolabeled leptin into the brain has been described [44]. Future studies are required to elucidate the mechanisms of leptin transport into the CNS, as understanding such mechanisms is critical for an understanding of the complex responses regulated by leptin.

Hypothalamic Targets of Leptin

Several lines of evidence indicate that neuropeptide Y (NPY) neurons in the arcuate nucleus are targets of leptin. Expression of NPY mRNA in the arcuate is increased in response to fasting in normal rats and is markedly increased in *ob/ob* and *db/db* mice. Administration of leptin in *ob/ob* mice and fasted rats suppresses this NPY overexpression in the arcuate nucleus [9, 12]. Direct anatomic evidence of leptin acting on NPY neurons was provided by Mercer et al. [28] who colocalized leptin receptor and NPY mRNAs in neurons in the arcuate. Recently, Erickson et al. [45] found that *ob/ob* mice that also lacked the NPY gene had an attenuated phenotype with decreased food intake and body weight when compared to the *ob/ob* mutation alone. These findings suggest that the arcuate NPY neuron is an important target of circulating leptin.

However, arcuate NPY neurons do not appear to be the only target of circulating leptin. For example, although lack of NPY did attenuate the *ob/ob* phenotype, these mice were still overweight and had neuroendocrine abnormalities when compared to littermate controls. Additionally, NPY deficient mice respond to exogenous leptin, normally regulate feeding behavior, body weight, neuroendocrine status and have an unaltered physiological response to starvation [46, 47]. More persuasively from an anatomic perspective, long-form leptin receptors are found in other hypothalamic nuclear groups including the DMH, VMH and ventral premamillary nuclei. Undoubtedly, the action of leptin on NPY neurons in the arcuate nucleus is an important component of the physiological effects of circulating leptin. However, when taken together these findings indicate that the arcuate nucleus and its NPY neurons are not the sole targets of peripheral leptin. Moreover, these studies suggest that alternative central pathways of leptin action exist.

Another recently identified target are proopiomelanocortin (POMC) neurons in the arcuate nucleus. POMC and leptin receptor mRNA colocalize within the arcuate nucleus [33]. Moreover, leptin regulates the levels of POMC mRNA as *ob/ob* mice and food restricted rats have lowered levels of POMC compared in controls. Repletion of leptin elevates arcuate POMC levels to those of controls [48–50]. The specific product of the POMC gene that is important in mediating the effects of leptin has not yet been directly demonstrated. However, one likely candidate is α-melanin stimulating hormone (α-MSH), which is an agonist at melanocortin receptors [51]. Recently, a novel endogenous melanocortin antagonist, agouti-related peptide (AgRP), has been identified in the arcuate nucleus of the hypothalamus [52, 53]. Interestingly, leptin's acute effects on feeding can be attenuated by blocking melanocortin receptors [54]. Therefore, some of leptin's effects are likely due to the effects of POMC (α-MSH) and AgRP neurons acting on subpopulations of neurons containing melanocortin receptors.

Several pieces of evidence point to a critical role of the melanocortin 4 receptor (MC4-R) system in the regulation of body weight. First, the lethal yellow (A^y) mouse has an ectopic overexpression of the MC4-R antagonist, agouti protein which is normally expressed only in hair follicles and not in the brain [51]. These mice have an adult onset obesity with leptin resistance. In addition, targeted deletion of the MC4-R results in an obesity syndrome indistinguishable from the A^y mouse [55]. Finally, transgenic overexpression of AgRP, and antagonism of MC4-Rs, also results in obesity [53]. Taken together, these findings suggest that central MC4-R populations regulate body weight and metabolism. Neurons containing MC4-Rs localize to the paraventricular hypothalamic nucleus, dorsomedial hypothalamic nucleus, and the lateral hypothalamic area [56]. However, the specific populations of neurons responsible for this and the efferent connections of these neurons are not well understood.

Leptin-Activated Cell Groups

The distribution of long form leptin receptors in the hypothalamus suggests that signaling via these receptors underlies leptin's physiological effects. Direct functional evidence of leptin action through hypothalamic receptors was provided by Vaisse and colleagues. This study demonstrated activation of Stat3 protein in the hypothalamus following systemic administration of leptin in normal mice but not in leptin-insensitive *db/db* mice [57]. Subsequently, activation of immediate early genes by leptin in specific nuclear groups was reported. These studies used expression of c-fos as a marker of neuronal activation. Fos, the protein product of the c-fos gene, binds to Jun proteins to form the AP-1 transcription factor. Although many functions of Fos in the CNS are still unknown, several studies have used the expression of Fos protein as a marker of cellular activation in the CNS in response to changes in neuroendocrine status [58]. An initial study reported Fos-activation in the paraventricular nucleus but not in other hypothalamic sites after intraperitoneal (i.p.) leptin administration in *ob/ob* mice [59]. Van Dijk et al. [60] found Fos-IR in the paraventricular and dorsomedial hypothalamic nuclei and in the central nucleus of the amygdala after intracerebroventricular (ICV) injection of leptin. However, it is still unclear how leptin, which is secreted by peripheral white adipose tissue, reaches the brain as circulating leptin is thought to be excluded by the blood-brain barrier.

Alternatively, we chose to use a fundamentally different model and investigated the CNS sites that are activated by intravenous (i.v.) leptin. Our results demonstrated that i.v. administration of leptin to fed rats activates several nuclear groups in the rat brain thought to be involved in regulation of energy balance including the ventromedial (VMH), dorsomedial (DMH), and paraventricular (PVH) hypothalamic nuclei (fig. 2, 3) [61]. Although little has been reported regarding the physiological effects of different doses of leptin in rats, the 1.0-mg/kg dose administered twice daily blunts starvation-induced changes in ACTH, corticosterone, thyroxine, hypothalamic NPY mRNA, and estrus delay in mice [6]. To date, the responses of rats in the fasted state have not been examined and directly compared to the distributions of Fos-IR in the fed state.

The activation of cells in VMH and DMH is consistent with a large body of literature that implicates this region in feeding and nutritional regulation. Large lesions of the VMH produce hyperphagia and obesity [62]. Later studies showed that hyperphagia and obesity could be produced by more dorsal lesions involving the PVH [63]. Lesions of the DMH result in complex effects on long term growth and body composition [64]. Our observations demonstrate populations of leptin-sensitive neurons in the dorsomedial VMH, caudal DMH, and PVH. Cells in one or all of these groups may be involved in producing the classic 'ventromedial nucleus' syndrome.

We did not find leptin-activated neurons in the medial subdivisions of the arcuate nucleus of the hypothalamus. This may seem surprising as this region has been reported to contain dense concentrations of leptin receptors. The arcuate nucleus also contains neuropeptide Y (NPY) neurons that project to the PVH [65], and are believed to increase feeding, particularly of carbohydrates [66]. However, since NPY is thought to be an excitatory neuromodulator in this pathway, we would expect that these neurons would be inhibited by leptin [9, 12, 67, 68] and thus would not express Fos-IR [69]. However, we did find leptin activated cells in the lateral edges of the arcuate (fig. 2E). This region of the hypothalamus also contains neurons containing proopiomelanocortin (POMC) and growth hormone releasing hormone [65, 70]. Activation of both of these types of neurons is consistent with in vivo data suggesting that leptin positively regulates POMC expression [48–50] and growth hormone secretion [14].

Administration of leptin also induced Fos-IR in the superior lateral subdivision of the parabrachial nucleus in the pons (fig. 3E) [61]. Interestingly, neurons in this nucleus contain cholecystokinin (CCK) and project to the ventrobasal hypothalamus including the VMH and DMH [71]. CCK is a

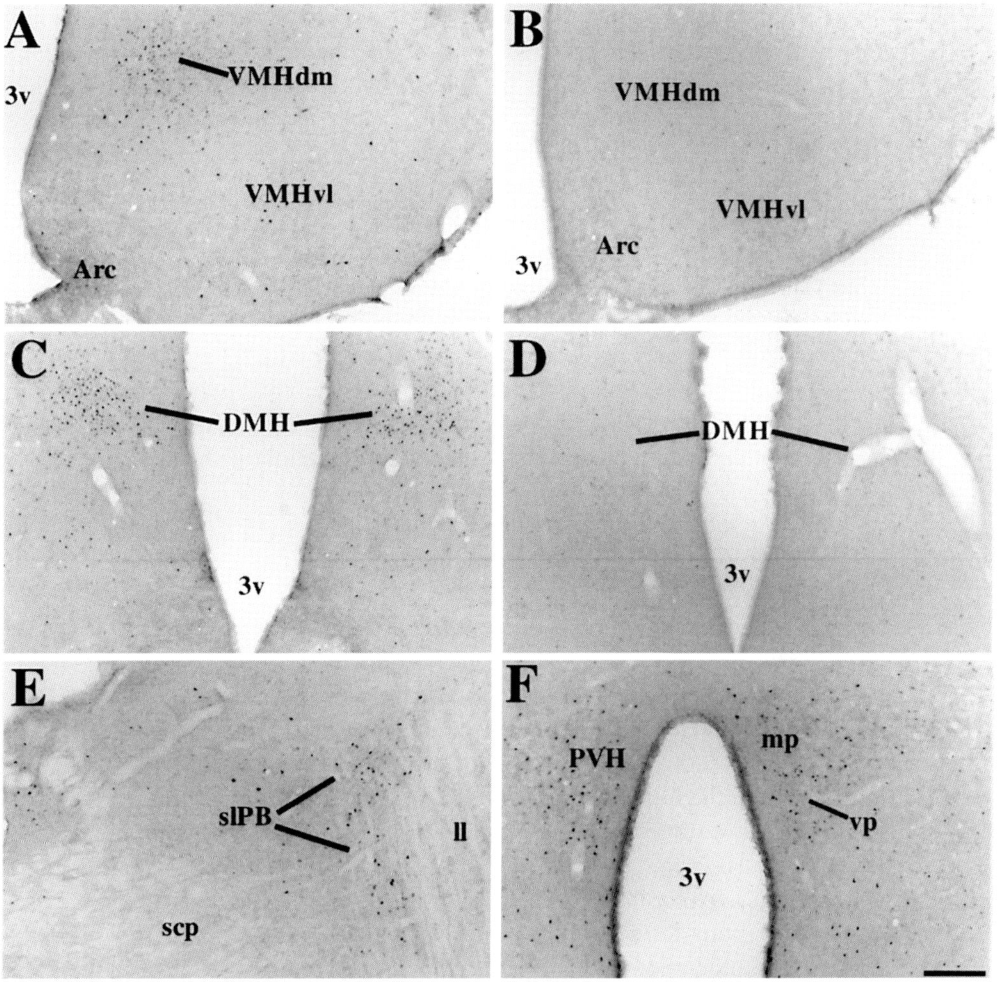

Fig. 3. Photomicrographs demonstrating the distribution of Fos-like immunoreactivity (Fos-IR) 2 h following intravenous administration of leptin (1.0 mg/kg; *A, C, E–F*) or pyrogen-free saline (PFS; *B, D*). *A* At this mid-tuberal level of the hypothalamus, Fos-IR is observed in the dorsomedial part of the ventromedial hypothalamic nucleus (VMHdm) but not in the ventrolateral VMH (VMHvl) or in the arcuate nucleus (Arc). *B* Very few cells are seen in the VMHdm following PFS. *C* Prominent Fos-IR is seen in the caudal dorsomedial hypothalamic nucleus (DMH) but not in the caudal Arc. Very little of the VMH is still present at this caudal tuberal level of the hypothalamus. *D* Few immunoreactive cells are seen in the DMH following PFS. *E* Fos-IR is observed in the superior lateral parabrachial subnucleus (slPB). *F* Fos-IR is observed in ventral parvicellular (vp) subdivision of the paraventricular nucleus of the hypothalamus (PVH) 2 h following leptin administration. ll = Lateral lemniscus; mp = medial parvicellular subdivision of the PVH; scp = superior cerebellar peduncle; 3v = third ventricle. Bar = 200 μm. Reprinted with permission from Elmquist et al. [61].

potent inhibitor of food intake and also regulates a number of endocrine and autonomic functions [72]. The ability of CCK and leptin to act in a synergistic fashion to inhibit food intake [73] suggests that there is a functional interaction of the pathways underlying responses to peripheral leptin and CCK. Although the neuroanatomic basis for this response remains to be established, it is plausible that leptin may directly or indirectly increase the sensitivity of CCKergic pathways that inhibit feeding.

Recently, a leptin inducible inhibitor of leptin-signal transduction has been described [74]. This inhibitor, a member of the suppressors of cytokine signalling (SOCS) family [75–77], named SOCS3, blocks leptin-induced activation of Stat3 in vitro. Moreover, the mRNA for SOCS3 is rapidly induced in the hypothalamus de novo following systemic leptin administration [74]. Therefore, this molecule may provide a more direct and relevant marker than Fos (or other immediate early genes) for leptin action in the brain.

Leptin Activates PVH-Projecting Neurons

The multiple effects of leptin on the autonomic nervous system and several neuroendocrine systems including the hypothalamo-pituitary-adrenal (HPA) and thyroid axes suggests that leptin's action on the paraventricular nucleus of the hypothalamus (PVH) is essential in mediating several of the effects of circulating leptin. The PVH is ideally positioned anatomically to regulate responses in the face of changing energy availability as it possesses chemically and anatomically specific projections to autonomic and endocrine control sites involved in maintenance of homeostasis [78, 79]. Previous anatomic and physiological experiments indicate that the PVH is critical in changing endocrine secretions following acute stressors. For example, the neuroendocrine parvicellular neurons (which project to the median eminence) regulate secretion of several hormones including TSH, growth hormone, and ACTH (all of which are affected by an absolute lack of leptin).

The PVH has also been implicated in the regulation of feeding behavior. For example, lesions of the PVH induce hyperphagia and obesity [63]. In addition, specific neurotransmitters and neuromodulators (including NPY) that innervate the PVH have profound effects on feeding behavior [66]. Finally, the PVH contains neurons in the dorsal, ventral, and lateral parvicellular subdivisions that directly innervate parasympathetic and sympathetic preganglionic neurons in the medulla and spinal cord [78]. These projections give the hypothalamus a direct input to the autonomic nervous system. Therefore the anatomic makeup of the PVH makes it ideally positioned to act as a final common pathway in regulating the marked responses to circulating leptin.

Leptin treatment induces Fos-IR within the parvicellular autonomic subnuclei in the PVH (fig. 3F). These neurons are a major source of descending axons to autonomic preganglionic neurons [78, 79] within the medulla and spinal cord. For example, stimulation of oxytocin containing neurons in the parvicellular subdivisions of the PVH that project to the dorsal motor nucleus of the vagus and the nucleus ambiguus (parasympathetic preganglionic neurons) stimulates gastric acid secretion and induces bradycardia [80]. In addition, as the dorsal motor nucleus of the vagus is the source of parasympathetic innervation to the pancreas, the PVH is positioned to regulate insulin secretion. Finally, the PVH directly innervates sympathetic preganglionic neurons that affect heart rate, blood pressure, and thermogenesis [79]. Activation of autonomic projecting neurons by leptin indicates a potential mechanism for leptin to regulate diverse processes such as thermoregulation, including that mediated by brown adipose tissue, insulin secretion, metabolic rate, and cardiovascular status. This is relevant as leptin administration increases sympathetic nerve activity [81] and norepinephrine turnover in brown adipose tissue [82].

The activation of the PVH by leptin could occur directly as low levels of leptin receptors are present in the PVH [32]. However, leptin receptors are found in the highest densities in the ventrobasal hypothalamus in regions known to project to the PVH. Therefore it is plausible that leptin activation of the PVH is due to innervation from leptin responsive neurons. To test this hypothesis directly, we investigated the distribution of leptin-activated projections (Fos-containing cells) to the PVH [83]. This was accomplished using stereotaxic injections of the retrograde tracer, cholera toxin B subunit (CTb) into the PVH followed by injections of leptin 5–7 days later. We found that many neurons in the dorsomedial hypothalamic nucleus (DMH), express Fos after systemic administration of leptin, and a very high percentage of these neurons project directly to the PVH (fig. 2) [83]. We have not yet identified the chemical nature of these leptin-activated projections.

The DMH and Leptin Responses

The dense distribution of leptin receptors in the DMH suggests that this nuclear group may be involved in coordinating the central response to leptin. The DMH lies caudal to the PVH and dorsal to the ventromedial hypothalamic nucleus (VMH). The DMH has been implicated in the regulation of a wide range of physiological processes including ingestive behavior, insulin secretion, cardiovascular function, and neuroendocrine status. Lesions of the DMH result in changes in pancreatic nerve activity [84] while stimulation of the DMH causes hyperglycemia and increases in plasma catecholamines [85].

These findings suggest a role of the DMH in regulating insulin secretion, presumably through interactions with the parasympathetic (dorsal motor nucleus of the vagus) and sympathetic (intermediolateral cell column of the spinal cord) preganglionic neurons. Additionally, lesions of the DMH cause alterations in feeding behavior and complex effects on long-term growth and body composition [64].

The DMH has extensive intrahypothalamic projections and a smaller but significant projection to the dorsal motor nucleus of the vagus in the medulla (pancreatic parasympathetic neurons) [86]. A recent anterograde tracing study by Swanson and colleagues confirmed the massive intrahypothalamic projection patterns of the DMH, but did not find a significant projection to the medulla [87]. The lack of labeling in parasympathetic and sympathetic preganglionic neurons indicates that the alterations of autonomic function elicited by DMH stimulation are not due to direct projections. However, a major target of DMH efferents is the PVH. This projection gives the DMH indirect inputs to the autonomic preganglionic neurons [78, 79]. Specifically, the DMH targets the dorsal, ventral, and lateral parvicellular PVH subdivisions that directly innervate parasympathetic and sympathetic preganglionic neurons in the medulla and spinal cord. Consistent with this suggestion, a transneuronal labeling study using injections of pseudorabies virus into the heart transsynaptically labeled neurons in the DMH [88]. The DMH has been implicated in the regulation of insulin secretion, feeding behavior, thermoregulation, and cardiovascular control. In addition, the DMH contains leptin receptors, expresses SOCS-3 mRNA and Fos-IR following leptin administration, and heavily innervates the PVH [32, 61, 83]. Therefore, it is plausible that leptin activation of this nuclear group results in changes in key autonomic and endocrine parameters including body weight, neuroendocrine status, insulin and glucose levels, blood pressure, and body temperature (fig. 4).

The VMH and Leptin Responses

Interestingly, relatively few cells observed in the VMH contained both Fos-IR and CTb following leptin administration and CTb injections in the PVH. This observation suggests that the leptin-activated cells in the VMH project to sites other than the PVH. Consistent with this hypothesis, when the CTb injections were centered in the subparaventricular zone, many double-labeled cells were seen in the dorsomedial VMH (fig. 4). The subparaventricular zone lies in the anterior hypothalamic area ventral and anterior to the PVH and receives a very dense innervation from the suprachiasmatic nucleus, the circadian pacemaker of the mammalian brain [89–91]. The VMH has been

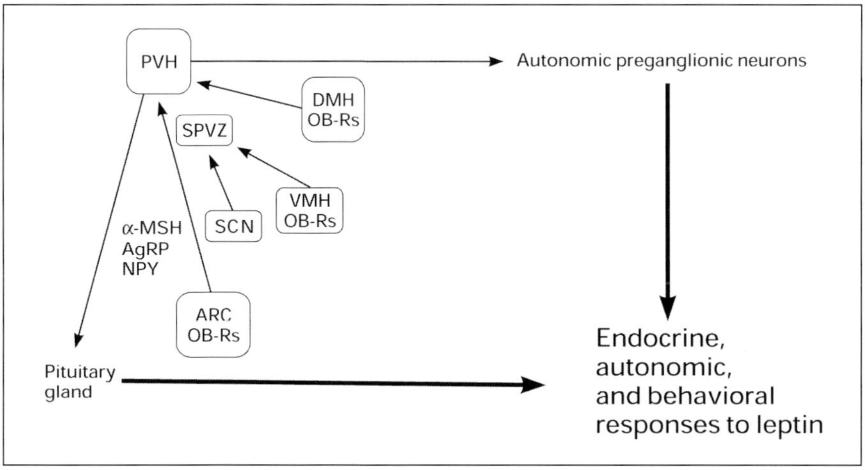

Fig. 4. A schematic drawing of the rat brain in sagittal section demonstrating a neuroanatomical model of leptin action: Circulating leptin acts on cell groups containing leptin receptors (OBRs) within the arcuate (Arc), dorsomedial (DMH), and ventromedial (VMH) hypothalamic nuclei. Ultimately, activation of the autonomic and neuroendocrine components of the paraventricular hypothalamic nucleus (PVH) is responsible for the physiological effects of leptin. We hypothesize that intravenous leptin inhibits Arc neuropeptide Y (NPY) and agouti-related peptide (AgRP) neurons and activates α-melanocyte stimulating hormone (α-MSH) neurons that innervate the PVH. Simultaneously, circulating leptin activates neurons in the DMH whose efferent projections converge on PVH neurons that innervate sympathetic and parasympathetic preganglionic neurons in the medulla and spinal cord. Additionally, circulating leptin activates neurons in the VMH whose efferent projections converge on the subparaventricular zone (SPVZ). The SPVZ also receives dense innervation from the suprachiasmatic nucleus (SCN). Engagement of these parallel pathways is responsible for the manifestation of the physiological effects of circulating leptin. Modified with permission from Elmquist et al. [83].

reported to regulate feeding and metabolism as lesions of the VMH and surrounding fiber pathways result in hyperphagia, obesity, diabetes, and elevated levels of corticosteroids. The anatomic makeup of the region (inputs from the VMH and the suprachiasmatic nucleus), suggests that the subparaventricular zone is essential in integrating nutritional and circadian information into endocrine responses [89–91].

Food restriction in the dark period results in changes in the circadian rhythms of corticosterone, illustrating a link between nutritional signals and corticosterone secretion [92–94]. Lesions of the VMH, but not the suprachiasmatic nucleus, abolish this change in circadian corticosterone levels in food-restricted rats [93, 95]. Additionally, lesions or inhibition of the VMH alter diurnal corticosterone rhythms [96–98]. These findings suggest that the VMH may be critical in linking nutritional status to circadian neuroendocrine responses. Interestingly, an inverse relationship of plasma leptin and corticosteroid levels exist in mice and humans [6, 99]. Food restriction reverses the phase of both the corticosterone and leptin circadian rhythms such that they maintain their inverse relationship [100]. Finally, the dorsomedial VMH contains leptin receptors [12, 29, 32], and is activated by i.v. leptin administration [61, 83]. These findings suggest that populations of leptin-activated neurons in the VMH innervate the subparaventricular zone, providing an anatomic substrate through which leptin may regulate the secretion of hormones such as corticosterone across the circadian cycle. Moreover, the subparaventricular zone projects densely to the DMH and not to the PVH [91], so the effects of leptin-activated neurons in the VMH that innervate the subparaventricular zone may ultimately be mediated by projections of the DMH to the PVH (fig. 4).

Conclusions and Future Directions

It is clear that leptin interacts with hypothalamic systems that regulate endocrine and autonomic function and the specific neuroanatomic pathways underlying these responses are now being identified. The coupling of neuroanatomic techniques with new genetic and molecular tools is essential in elucidating the mechanisms of body weight regulation and energy homeostasis. Results from several studies suggest that a discrete set of hypothalamic pathways may underlie leptin's autonomic, endocrine, and behavioral effects. We hypothesize that multiple pathways are influenced by circulating leptin and that these pathways acting in concert underlie the multiple CNS responses to circulating leptin (fig. 4). Future studies will concentrate on the outputs of leptin responsive neurons in the hypothalamus to effector sites such as the autonomic preganglionic neurons and the cerebral cortex. A stepwise elucidation of the anatomic pathways engaged by leptin will enhance our understanding of the physiological significance of this fundamental hormone.

Acknowledgment

This work was supported by USPHS grants MH56537 and DK53301.

References

1 Zhang Y, Proenca R, Maffei M, Barone M, Leopold L, Friedman JM: Positional cloning of the mouse obese gene and its human homologue. Nature 1994;372:425–432.

2 Spiegelman BM, Flier JS: Adipogenesis and obesity: Rounding out the big picture. Cell 1996;87: 377–389.

3 Halaas JL, Gajiwala KS, Maffei M, Cohen SL, Chait BT, Rabinowitz D, Lallone RL, Burley SK, Friedman JM: Weight-reducing effects of the plasma protein encoded by the obese gene. Science 1995;269:543–546.

4 Campfield LA, Smith FJ, Guisez Y, Devos R, Burn P: Recombinant mouse OB protein: Evidence for a peripheral signal linking adiposity and central neural networks. Science 1995;269:546–549.

5 Pelleymounter MA, Cullen MJ, Baker MB, Hecht R, Winters D, Boone T, Collins F: Effects of the obese gene product on body weight regulation in ob/ob mice. Science 1995;269:540–543.

6 Ahima RS, Prabakaran D, Mantzoros C, Qu D, Lowell B, Maratos-Flier E, Flier JS: Role of leptin in the neuroendocrine response to fasting. Nature 1996;382:250–252.

7 Frederich RC, Hamann A, Anderson S, Lollmann B, Lowell BB, Flier JS: Leptin levels reflect body lipid content in mice: Evidence for diet-induced resistance to leptin action. Nat Med 1995;1:1311–1314.

8 Schwartz MW, Dallman MF, Woods SC: Hypothalamic response to starvation: Implications for the study of wasting disorders. Am J Physiol 1995;269:R949–R957.

9 Stephens TW, Basinski M, Bristow PK, Bue-Valleskey JM, Burgett SG, Craft L, Hale J, Hoffmann J, Hsiung HM, Kriauciunas A, et al: The role of neuropeptide Y in the antiobesity action of the obese gene product. Nature 1995;377:530–532.

10 Heiman ML, Ahima RS, Craft LS, Schoner B, Stephens TW, Flier JS: Leptin inhibition of the hypothalamic-pituitary-adrenal axis in response to stress. Endocrinology 1997;138:3859–3863.

11 Huang Q, Rivest R, Richard D: Effects of leptin on corticotropin-releasing factor (CRF) synthesis and CRF neuron activation in the paraventricular hypothalamic nucleus of obese (ob/ob) mice. Endocrinology 1998;139:1524–1532.

12 Schwartz MW, Seeley RJ, Campfield LA, Burn P, Baskin DG: Identification of targets of leptin action in rat hypothalamus. J Clin Invest 1996;98:1101–1106.

13 Sawchenko PE, Brown ER, Chan RK, Ericsson A, Li HY, Roland BL, Kovacs KJ: The paraventricular nucleus of the hypothalamus and the functional neuroanatomy of visceromotor responses to stress. Prog Brain Res 1996;107:201–222.

14 Carro E, Senaris R, Considine RV, Casanueva FF, Dieguez C: Regulation of in vivo growth hormone secretion by leptin. Endocrinology 1997;138:2203–2206.

15 Quintela M, Senaris R, Heiman ML, Casanueva FF, Dieguez C: Leptin inhibits in vitro hypothalamic somatostatin secretion and somatostatin mRNA levels. Endocrinology 1997;138:5641–5644.

16 Blake NG, Eckland DJ, Foster OJ, Lightman SL: Inhibition of hypothalamic thyrotropin-releasing hormone messenger ribonucleic acid during food deprivation. Endocrinology 1991;129:2714–2718.

17 Legradi G, Emerson CH, Ahima RS, Flier JS, Lechan RM: Leptin prevents fasting-induced suppression of prothyrotropin-releasing hormone messenger ribonucleic acid in neurons of the hypothalamic paraventricular nucleus. Endocrinology 1997;138:2569–2576.

18 Chehab FF, Lim ME, Lu R: Correction of the sterility defect in homozygous obese female mice by treatment with the human recombinant leptin. Nat Genet 1996;12:318–320.

19 Barash IA, Cheung CC, Weigle DS, Ren H, Kabigting EB, Kuijper JL, Clifton DK, Steiner RA: Leptin is a metabolic signal to the reproductive system. Endocrinology 1996;137:3144–3147.

20 Ahima RS, Dushay J, Flier SN, Prabakaran D, Flier JS: Leptin accelerates the onset of puberty in normal female mice. J Clin Invest 1997;99:391–395.

21 Cheung CC, Thornton JE, Kuijper JL, Weigle DS, Clifton DK, Steiner RA: Leptin is a metabolic gate for the onset of puberty in the female rat. Endocrinology 1997;138:855–858.

22 Chehab FF, Mounzih K, Lu R, Lim ME: Early onset of reproductive function in normal female mice treated with leptin. Science 1997;275:88–90.

23 Mantzoros CS, Flier JS, Rogol AD: A longitudinal assessment of hormonal and physical alterations during normal puberty in boys. V. Rising leptin levels may signal the onset of puberty. J Clin Endocrinol Metab 1997;82:1066–1070.

24 Plant TM, Durrant AR: Circulating leptin does not appear to provide a signal for triggering the initiation of puberty in the male rhesus monkey (*Macaca mulatta*). Endocrinology 1997;138: 4505–4508.
25 Yu WH, Kimura M, Walczewska A, Karanth S, McCann SM: Role of leptin in hypothalamic-pituitary function. Proc Natl Acad Sci USA 1997;94:1023–1028.
26 Montague CT, Farooqi IS, Whitehead JP, Soos MA, Rau H, Wareham NJ, Sewter CP, Digby JE, Mohammed SN, Hurst JA, Cheetham CH, Earley AR, Barnett AH, Prins JB, O'Rahilly S: Congenital leptin deficiency is associated with severe early-onset obesity in humans. Nature 1997;387:903–908.
27 Tartaglia LA, Dembski M, Weng X, Deng N, Culpepper J, Devos R, Richards GJ, Campfield LA, Clark FT, Deeds J, et al: Identification and expression cloning of a leptin receptor, OB-R. Cell 1995;83:1263–1271.
28 Mercer JG, Hoggard N, Williams LM, Lawrence CB, Hannah LT, Morgan PJ, Trayhurn P: Coexpression of leptin receptor and prepreoneuropeptide Y mRNA in arcuate nucleus of mouse hypothalamus. J Neuroendocrinol 1996;8:733–735.
29 Mercer JG, Hoggard N, Williams LM, Lawrence CB, Hannah LT, Trayhurn PL: Localization of leptin receptor mRNA and the long form splice variant (Ob-Rb) in mouse hypothalamus and adjacent brain regions by in situ hybridization. FEBS Lett 1996;387:113–116.
30 Mercer JG, Moar KM, Rayner DV, Trayhurn P, Hoggard N: Regulation of leptin receptor and NPY gene expression in hypothalamus of leptin-treated obese (ob/ob) and cold-exposed lean mice. FEBS Lett 1997;402:185–188.
31 Fei H, Okano HJ, Li C, Lee GH, Zhao C, Darnell R, Friedman JM: Anatomic localization of alternatively spliced leptin receptors (Ob-R) in mouse brain and other tissues. Proc Natl Acad Sci USA 1997;94:7001–7005.
32 Elmquist JK, Bjørbæk C, Ahima RS, Flier JS, Saper CB: Distributions of leptin receptor isoforms in the rat brain. J Comp Neurol 1998;395:535–547.
33 Cheung CC, Clifton DK, Steiner RA: Proopiomelanocortin neurons are direct targets for leptin in the hypothalamus. Endocrinology 1997;138:4489–4492.
34 Lee GH, Proenca R, Montez JM, Carroll KM, Darvishzadeh JG, Lee JI, Friedman JM: Abnormal splicing of the leptin receptor in diabetic mice. Nature 1996;379:632–635.
35 Chen H, Charlat O, Tartaglia LA, Woolf EA, Weng X, Ellis SJ, Lakey ND, Culpepper J, Moore KJ, Breitbart RE, Duyk GM, Tepper RI, Morgenstern JP: Evidence that the diabetes gene encodes the leptin receptor: Identification of a mutation in the leptin receptor gene in db/db mice. Cell 1996; 84:491–495.
36 White DW, Wang DW, Chua SC Jr, Morgenstern JP, Leibel RL, Baumann H, Tartaglia LA: Constitutive and impaired signaling of leptin receptors containing the Gln→Pro extracellular domain fatty mutation. Proc Natl Acad Sci USA 1997;94:10657–10662.
37 Savioz A, Charnay Y, Huguenin C, Graviou C, Greggio B, Bouras C: Expression of leptin receptor mRNA (long splice form variant) in the human cerebellum. NeuroReport 1997;8:3123–3126.
38 Mercer JG, Moar KM, Hoggard N: Localization of leptin receptor (Ob-R) messenger ribonucleic acid in the rodent hindbrain. Endocrinology 1998;139:29–34.
39 Couce ME, Burguera B, Parisi JE, Jensen MD, Lloyd RV: Localization of leptin receptor in the human brain. Neuroendocrinology 1997;66:145–150.
40 Hakansson ML, Brown H, Ghilardi N, Skoda RC, Meister B: Leptin receptor immunoreactivity in chemically defined target neurons of the hypothalamus. J Neurosci 1998;18:559–572.
41 Smedh U, Hakansson ML, Meister B, Uvnas-Moberg K: Leptin injected into the fourth ventricle inhibits gastric emptying (in process citation). NeuroReport 1998;9:297–301.
42 Bi S, Gavrilova O, Gong DW, Mason MM, Reitman M: Identification of a placental enhancer for the human leptin gene. J Biol Chem 1997;272:30583–30588.
43 Broadwell RD, Brightman MW: Entry of peroxidase into neurons of the central and peripheral nervous systems from extracerebral and cerebral blood. J Comp Neurol 1976;166:257–283.
44 Banks WA, Kastin AJ, Huang W, Jaspan JB, Maness LM: Leptin enters the brain by a saturable system independent of insulin. Peptides 1996;17:305–311.
45 Erickson JC, Hollopeter G, Palmiter RD: Attenuation of the obesity syndrome of ob/ob mice by the loss of neuropeptide Y. Science 1996;274:1704–1707.

46 Erickson JC, Clegg KE, Palmiter RD: Sensitivity to leptin and susceptibility to seizures of mice lacking neuropeptide Y (see comments). Nature 1996;381:415–421.
47 Erickson JC, Ahima RS, Hollopeter G, Flier JS, Palmiter RD: Endocrine function of neuropeptie Y knockout mice. Regul Pept 1997;70:199–202.
48 Thornton JE, Cheung CC, Clifton DK, Steiner RA: Regulation of hypothalamic proopiomelanocortin mRNA by leptin in ob/ob mice. Endocrinology 1997;138:5063–5066.
49 Schwartz MW, Seeley RJ, Woods SC, Weigle DS, Campfield LA, Burn P, Baskin DG: Leptin increases hypothalamic pro-opiomelanocortin mRNA expression in the rostral arcuate nucleus. Diabetes 1997;46:2119–2123.
50 Mizuno TM, Kleopoulos SP, Bergen HT, Roberts JL, Priest CA, Mobbs CV: Hypothalamic proopiomelanocortin mRNA is reduced by fasting in ob/ob and db/db mice, but is stimulated by leptin. Diabetes 1998;47:294–297.
51 Fan W, Boston BA, Kesterson RA, Hruby VJ, Cone RD: Role of melanocortinergic neurons in feeding and the agouti obesity syndrome. Nature 1997;385:165–168.
52 Shutter JR, Graham M, Kinsey AC, Scully S, Luthy R, Stark KL: Hypothalamic expression of ART, a novel gene related to agouti, is up-regulated in obese and diabetic mutant mice. Genes Dev 1997;11:593–602.
53 Ollmann MM, Wilson BD, Yang YK, Kerns JA, Chen Y, Gantz I, Barsh GS: Antagonism of central melanocortin receptors in vitro and in vivo by agouti-related protein. Science 1997;278: 135–138.
54 Seeley RJ, Yagaloff KA, Fisher SL, Burn P, Thiele TE, van Dijk G, Baskin DG, Schwartz MW: Melanocortin receptors in leptin effects. Nature 1997;390:349.
55 Huszar D, Lynch CA, Fairchild-Huntress V, Dunmore JH, Fang Q, Berkemeier LR, Gu W, Kesterson RA, Boston BA, Cone RD, Smith FJ, Campfield LA, Burn P, Lee F: Targeted disruption of the melanocortin-4 receptor results in obesity in mice. Cell 1997;88:131–141.
56 Mountjoy KG, Mortrud MT, Low MJ, Simerly RB, Cone RD: Localization of the melanocortin-4 receptor (MC4-R) in neuroendocrine and autonomic control circuits in the brain. Mol Endocrinol 1994;8:1298–1308.
57 Vaisse C, Halaas JL, Horvath CM, Darnell JE Jr, Stoffel M, Friedman JM: Leptin activation of Stat3 in the hypothalamus of wild-type and ob/ob mice but not db/db mice. Nat Genet 1996;14: 95–97.
58 Hoffman GE, Smith MS, Verbalis JG: c-Fos and related immediate early gene products as markers of activity in neuroendocrine systems. Front Neuroendocrinol 1993;14:173–213.
59 Woods AJ, Stock MJ: Leptin activation in hypothalamus (letter). Nature 1996;381:745.
60 Van Dijk G, Thiele TE, Donahey JC, Campfield LA, Smith FJ, Burn P, Bernstein IL, Woods SC, Seeley RJ: Central infusions of leptin and GLP-1-(7-36) amide differentially stimulate c-FLI in the rat brain. Am J Physiol 1996;271:R1096–R1100.
61 Elmquist JK, Ahima RS, Maratos-Flier E, Flier JS, Saper CB: Leptin activates neurons in ventrobasal hypothalamus and brainstem. Endocrinology 1997;138:839–842.
62 Hetherington AW, Ranson SW: Hypothalamic lesions and adiposity in the rat. Anat Rec 1940;78: 149.
63 Gold RM: Hypothalamic obesity: The myth of the ventromedial nucleus. Science 1973;182:488–490.
64 Bernardis LL, Bellinger LL: The dorsomedial hypothalamic nucleus revisited: 1986 update. Brain Res 1987;44:321–381.
65 Baker RA, Herkenham M: Arcuate nucleus neurons that project to the hypothalamic paraventricular nucleus: Neuropeptidergic identity and consequences of adrenalectomy on mRNA levels in the rat. J Comp Neurol 1995;358:518–530.
66 Leibowitz SF: Brain neuropeptide Y: An integrator of endocrine, metabolic and behavioral processes. Brain Res Bull 1991;27:333–337.
67 Spanswick D, Smith MA, Groppi VE, Logan SD, Ashford ML: Leptin inhibits hypothalamic neurons by activation of ATP-sensitive potassium channels. Nature 1997;390:521–525.
68 Glaum SR, Hara M, Bindokas VP, Lee CC, Polonsky KS, Bell GI, Miller RJ: Leptin, the obese gene product, rapidly modulates synaptic transmission in the hypothalamus. Mol Pharmacol 1996; 50:230–235.

69 Chan RK, Brown ER, Ericsson A, Kovacs KJ, Sawchenko PE: A comparison of two immediate-early genes, c-fos and NGFI-B, as markers for functional activation in stress-related neuroendocrine circitry. J Neurosci 1993;13:5126–5138.
70 Sawchenko PE, Swanson LW, Rivier J, Vale WW: The distribution of growth-hormone-releasing factor (GRF) immunoreactivity in the central nervous system of the rat: An immunohistochemical study using antisera directed against rat hypothalamic GRF. J Comp Neurol 1985;237:100–115.
71 Fulwiler CE, Saper CB: Cholecystokinin-immunoreactive innervation of the ventromedial hypothalamus in the rat: Possible substrate for autonomic regulation of feeding. Neurosci Lett 1985;53:289–296.
72 Weller A, Smith GP, Gibbs J: Endogenous cholecystokinin reduces feeding in young rats. Science 1990;247:1589–1591.
73 Barrachina MD, Martínez V, Wang L, Wei JY, Tach Y: Synergistic interaction between leptin and cholecystokinin to reduce short-term food intake in lean mice. Proc Natl Acad Sci USA 1997;94:10455–10460.
74 Bjørbæk C, Elmquist JK, Frazier JD, Shoelson SE, Flier JS: Identification of SOCS-3 as a potential mediator of central leptin resistance. Mol Cell 1998;1:619–625.
75 Endo TA, Masuhara M, Yokouchi M, Suzuki R, Sakamoto H, Mitsui K, Matsumoto A, Tanimura S, Ohtsubo M, Misawa H, Miyazaki T, Leonor N, Taniguchi T, Fujita T, Kanakura Y, Komiya S, Yoshimura A: A new protein containing an SH2 domain that inhibits JAK kinases. Nature 1997;387:921–924.
76 Starr R, Willson TA, Viney EM, Murray LJ, Rayner JR, Jenkins BJ, Gonda TJ, Alexander WS, Mitcalf D, Nicola NA, Hilton DJ: A family of cytokine-inducible inhibitors of signalling. Nature 1997;387:917–921.
77 Naka T, Narazaki M, Hirata M, Matsumoto T, Minamoto S, Aono A, Nishimoto N, Kajita T, Taga T, Yoshizaki K, Akira S, Kishimoto T: Structure and function of a new STAT-induced STAT inhibitor. Nature 1997;387:924–929.
78 Swanson LW, Sawchenko PE: Hypothalamic integration: Organization of the paraventricular and supraoptic nuclei. Annu Rev Neurosci 1983;6:269–324.
79 Saper CB: Central autonomic system; in Paxinos G (ed): The Rat Nervous System. San Diego, Academic Press, 1995, pp 107–135.
80 Rogers RC, Hermann GE: Dorsal medullary oxytocin, vasopressin, oxytocin antagonist, and TRH effects on gastric acid secretion and heart rate. Peptides 1985;6:1143–1148.
81 Haynes WG, Morgan DA, Walsh SA, Mark AL, Sivitz WI: Receptor-mediated regional sympathetic nerve activation by leptin. J Clin Invest 1997;100:270–278.
82 Collins S, Kuhn CM, Petro AE, Swick AG, Chrunyk BA, Surwit RS: Role of leptin in fat regulation (letter). Nature 1996;380:677.
83 Elmquist JK, Ahima RS, Elias CF, Flier JS, Saper CB: Leptin activates distinct projections from the dorsomedial and ventromedial hypothalamic nuclei. Proc Natl Acad USA 1998;95:741–746.
84 Yoshimatsu H, Niijima A, Oomura Y, Yamabe K, Katafuchi T: Effects of hypothalamic lesion on pancreatic autonomic nerve activity in the rat. Brain Res 1984;303:147–152.
85 Frohman LA, Bernardis LL: Effect of hypothalamic stimulation on plasma glucose, insulin, and glucagon levels. Am J Physiol 1971;221:1596–1603.
86 ter Horst GJ, Luiten PG: The projections of the dorsomedial hypothalamic nucleus in the rat. Brain Res Bull 1986;16:231–248.
87 Thompson RH, Canteras NS, Swanson LW: Organization of projections from the dorsomedial nucleus of the hypothalamus: A PHA-L study in the rat. J Comp Neurol 1996;376:143–173.
88 ter Horst GJ, Hautvast RW, De Jongste MJ, Korf J: Neuroanatomy of cardiac activity-regulating circuitry: A transneuronal retrograde viral labelling study in the rat. Eur J Neurosci 1996;8:2029–2041.
89 Watts AG, Swanson LW: Efferent projections of the suprachiasmatic nucleus. II. Studies using retrograde transport of fluorescent dyes and simultaneous peptide immunohistochemistry in the rat. J Comp Neurol 1987;258:230–252.

90 Watts AG, Swanson LW, Sanchez-Watts G: Efferent projections of the suprachiasmatic nucleus. I. Studies using anterograde transport of *Phaseolus vulgaris* leucoagglutinin in the rat. J Comp Neurol 1987;258:204–229.
91 Watts AG: The efferent projections of the suprachiasmatic nucleus: Anatomical insights into the control of circadian rhythms; in Klein D, Moore RY, Reppert SM (eds): Suprachiasmatic Nucleus: The Mind's Clock. Oxford, Oxford University Press, 1991, pp 75–104.
92 Krieger DT: Food and water restriction shifts corticosterone, temperature, activity and brain amine periodicity. Endocrinology 1974;95:1195–1201.
93 Krieger DT, Hauser H, Krey LC: Suprachiasmatic nuclear lesions do not abolish food-shifted circadian adrenal and temperature rhythmicity. Science 1977;197:398–399.
94 Moberg GP, Bellinger LL, Mendel VE: Effect of meal feeding on daily rhythms of plasma corticosterone and growth hormone in the rat. Neuroendocrinology 1975;19:160–169.
95 Krieger DT: Ventromedial hypothalamic lesions abolish food-shifted circadian adrenal and temperature rhythmicity. Endocrinology 1980;106:649–654.
96 Bellinger LL, Bernardis LL, Mendel VE: Effect of ventromedial and dorsomedial hypothalamic lesions on circadian corticosterone rhythms. Neuroendocrinology 1976;22:216–225.
97 Choi S, Horsley C, Aguila S, Dallman MF: The hypothalamic ventromedial nuclei couple activity in the hypothalamo-pituitary-adrenal axis to the morning fed or fasted state. J Neurosci 1996;16: 8170–8180.
98 Suemaru S, Darlington DN, Akana SF, Cascio CS, Dallman MF: Ventromedial hypothalamic lesions inhibit corticosteroid feedback reglation of basal ACTH during the trough of the circadian rhythm. Neuroendocrinology 1995;61:453–463.
99 Licinio J, Mantzoros C, Negrao AB, Cizza G, Wong ML, Bongiorno PB, Chrousos GP, Karp B, Allen C, Flier JS, Gold PW: Human leptin levels are pulsatile and inversely related to pituitary-adrenal function. Nat Med 1997;3:575–579.
100 Ahima RS, Prabakaran D, Flier JS: Postnatal leptin surge and regulation of circadian rhythm of leptin by feeding: Implications for energy homeostasis and neuroendocrine function. J Clin Invest 1998;101:1020–1027.

Joel K. Elmquist, DVM, PhD, Division of Endocrinology, Beth Israel Deaconess Medical Center, 325 Research North, 99 Brookline Avenue, Boston, MA 02215 (USA)
Tel. +1 617 667 0845, Fax +1 617 667 2927, E-Mail jelmquis@caregroup.harvard.edu

Ur E (ed): Neuroendocrinology of Leptin.
Front Horm Res. Basel, Karger, 2000, vol 26, pp 42–56

Leptin and the Neuroendocrinology of Fasting

Rexford S. Ahima

Department of Medicine, Division of Endocrinology, Beth Israel Deaconess Medical Center and Harvard Medical School, Boston, Mass., USA

Humans and other mammals are faced with the challenge of ensuring a constant supply of energy for cellular metabolism. Since food intake is limited by sleep-wake cycles and food shortages occur not infrequently, the ability to store excess calories and limit energy expenditure promotes survival. Mammals consume more calories at each meal than are required to meet immediate metabolic needs and store the excess energy as glycogen, fat and protein. The threat of fuel depletion during fasting triggers compensatory metabolic, endocrine and behavioral responses to limit energy expenditure and stimulate food intake. Classic studies carried out several decades ago provided the foundation of our current understanding of the relationship between fuel homeostasis and the endocrine system during fasting [1, 2]. As blood glucose levels fall in the postabsorptive and early phases of fasting there is a corresponding decrease in insulin and a rise in counterregulatory hormones such glucagon, epinephrine and cortisol, leading to stimulation of glucose production from hepatic glycogen stores for use by the brain, muscle and kidneys. With prolongation of the duration of fasting low insulin levels stimulate lipolysis and ketone body production. The switch to fat-based metabolism provides fatty acids and ketone bodies for metabolism by skeletal muscle and the brain, respectively. The net effect of the metabolic response to fasting mediated by lack of insulin is to supply alternate efficient fuels for metabolism, while reducing the need for glucose, and thereby preventing protein catabolism and protecting lean mass.

Fasting is also associated with profound alterations of autonomic and neuroendocrine function including suppression of sympathetic nervous activity, activation of the hypothalamic-pituitary-adrenal axis and suppression of reproductive and thyroid hormones [3–8]. Growth hormone levels are also altered depending on the duration of fasting and the species. For example, growth hormone falls with fasting in rodents but is elevated in the early phase of fasting in humans [1, 9]. In contrast, levels of insulin-like growth factor-I are suppressed during fasting [10]. These diverse changes in hormone levels are likely to promote survival by stimulating food ingestion and conserving energy through reduction of thyroid thermogenesis, and decreased growth and reproduction.

The signals which mediate the interaction between energy stores and diverse hormonal changes during fasting are largely unknown. Levels of the recently discovered hormone, leptin, are directly proportional to energy stores in adipose tissue [11–13] and decrease with fasting [14–16]. Since leptin deficiency and insensitivity to leptin are associated with endocrine abnormalities similar to fasting, we hypothesized that the fall in leptin is an important mediator of the adaptation to fasting [16]. Studies from our laboratory and others support such a role of leptin [16–18]. In this chapter, leptin along with other hormones which are regulated by nutritional status are proposed to inform the brain about declining energy levels and elicit integrated responses to counteract the effects of fasting.

Adipose Regulation and the Adaptation to Fasting

More than three decades ago Kennedy [19] proposed a model of energy homeostasis in which the amount of energy stored as fat in adipose tissue represented the balance between ingested calories and energy expenditure. In order for fat mass to remain stable over time, a surveillance system was thought to monitor changes in energy stores and elicit compensatory responses to match food intake to energy expenditure. This adipostatic model of energy balance was consistent with the observation that a decrease in energy stores from fasting or resection of adipose tissue resulted in hyperphagia and decreased energy expenditure [20]. Conversely, forced or involuntary feeding triggers a compensatory decrease in food intake in order to restore body weight and adiposity to a previous set point [20, 21].

The notion that signals related to energy stores could elicit compensatory responses to maintain energy balance received further support from studies by Coleman [22] using murine models of obesity. Using parabiosis experiments he determined that deficiency of a circulating anorectic factor in *ob/ob* mice led to hyperphagia and morbid obesity, and that insensitivity to this factor caused a similar syndrome in diabetes-prone *db/db* mice [22]. These observations have been confirmed by the recent cloning of the *obese* gene [23] and the demonstration that its protein product (named leptin) inhibits food intake and reduces body weight and adiposity [24–26]. The leptin receptor has also been cloned [27], and leptin insensitivity leading to hyperphagia and morbid obesity in *db/db* mice and Zucker rats has been localized to the *diabetes* and *fatty* mutations of the leptin receptor, respectively [28, 29].

Leptin is a protein hormone which is synthesized by adipose tissue in direct proportion to fat mass [11–13]. A rise in leptin levels with increasing adiposity is thought to act via a classic negative feedback loop to prevent obesity by inhibiting food intake and increasing thermogenesis [24–26]. The leptin receptor belongs to the family of class 1a cytokine receptors [30]. Although leptin receptor isoforms are present in several peripheral tissues, leptin is thought to regulate energy balance mainly by acting on neuronal targets in the hypothalamus [31–34]. In addition to exerting a more potent inhibition on food intake when administered directly into the brain [24–26], the view that leptin acts on targets in the brain to influence energy balance is supported by the existence of a saturable leptin transport system [35]. It has also been suggested that leptin receptors in the choroid plexus and endothelium may serve a transport function [27, 31, 36].

Although the view that leptin is an 'adipostatic hormone' whose dominant role is to prevent obesity is consistent with the development of hyperphagia, obesity and metabolic derangements in rodents and humans with leptin deficiency or insensitivity [24–26, 37–39], the role of leptin in the pathogenesis of obesity in normal animals is unclear. Lean and diet-induced obese rodents are less sensitive to leptin [24–26]. Additionally, hyperleptinemia in obese humans and rodents is thought to represent a state of leptin 'resistance' [12, 13]. However the mechanisms underlying the decreased sensitivity to leptin in normal animals are yet to be determined. Since the threat of starvation and fuel depletion is more common than the threat of obesity, it is more likely that starvation would constitute the dominant selective pressure for the development of mechanisms to ensure survival in the terrestial environment. Thus, the evolution of signals which are capable of detecting a decline in energy stores and stimulating compensatory responses to decrease energy expenditure and increase food intake would therefore be expected to confer greater survival advantage.

Insulin is a potential candidate as a signal linking energy stores to the brain during starvation. In addition to its well-known peripheral effects on fuel homeostasis [1], studies showing that insulin can be transported into the brain, reduce food intake, influence energy expenditure as well as regulate the expression of hypothalamic neuropeptides involved in energy balance have led to speculation that insulin has direct central actions [40, 41]. While this concept is consistent with a potential role of insulin as a central mediator of feeding and autonomic responses, it does not account for the diverse changes in thyroid, adrenal, gonadal and growth hormones during starvation. It is therefore likely that other factors related to energy stores may serve as signals to the brain during starvation.

Leptin as Sensor and Signal for the Adaptation to Starvation

Various aspects of the biology of leptin make it an ideal candidate as a mediator of the adaptation to starvation. First, leptin concentration is tightly coupled to adiposity and nutritional status [11–13] and can therefore serve as a sensor of long-term changes in energy stores. Second, since the fall in leptin levels in the early phase of fasting is rapid and out of proportion to the decrease in fat mass [14, 42], leptin could also serve as a sensor of short-term changes in energy stores. The latter role of leptin is consistent with the temporal relation between leptin and food intake [15, 16, 43]. Unlike insulin there is no meal-to-meal variation of leptin levels [43]. Rather, there is a diurnal rhythm of leptin with peak levels occurring at the end of the feeding cycle, i.e. the light cycle in rodents and dark cycle in humans, while the nadir of leptin precedes the onset of the feeding [15, 16, 43]. Prevention of nocturnal feeding blunts the leptin surge in rats [15], while a constant leptin infusion inhibits the expected increase in nocturnal feeding [44]. Third, leptin receptors are present in neuronal groups involved in feeding behavior and neuroendocrine function [31, 32, 34, 36]. Leptin receptor has been colocalized with various neuropeptides [45, 46], and leptin regulates the expression of neuropeptides implicated in feeding and hormonal regulation such as neuropeptide Y (NPY), proopiomelanocortin (POMC), corticotropin-releasing hormone and thyrotropin-releasing hormone [17, 47, 48].

The hypothesis that the fall in leptin may mediate the adaptation to fasting is also supported by similarities between autonomic, neuroendocrine and behavioral responses in leptin-deficient or -resistant rodents and starvation [1–3, 16, 47, 49–52]. Hypoleptinemia resulting from inherited disorders and starvation is associated with decreased sympathetic and thermoregulatory responses, increased glucocorticoid levels and suppression of growth and reproductive hormones. Similar neuroendocrine abnormalities have recently been observed in humans with inherited leptin deficiency and mutation of the leptin receptor [38, 39]. Although thyroid hormones are suppressed during fasting [2, 8, 17] and in humans with leptin receptor mutation [39], hypothyroidism has not been consistently demonstrated in *ob/ob* mice [49]. Since leptin treatment reverses metabolic as well as neuroendocrine abnormalities in *ob/ob* mice [24–26, 53–55], we reasoned that leptin deficiency or resistance was perceived as a state of continuous starvation. The development of hyperphagia, hypothermia, hypercorticism, hypothyroidism and hypogonadism in *ob/ob* mice could therefore represent compensatory metabolic and hormonal responses to starvation.

The hypothesis that the fall in leptin is a signal for the adaptation to starvation has been tested in normal mice [16] (table 1). Plasma leptin decreased rapidly by 50% within 24 h of initiation of fasting. Corticosterone and adrenocorticotropic hormone (ACTH) increased five- and twofold, respectively, while thyroxine levels fell. Leptin administration did not significantly alter corticosterone, ACTH and thyroxine during a 24-hour fast. Leptin concentration decreased further by 20% after 48 h of fasting, and was associated with some attenuation of the rise of corticosterone and ACTH and a further decrease in thyroxine levels (table 1). Leptin administration during a 48-hour fast blunted the activation of the pituitary-adrenal axis and prevented the fall in thyroxine and triiodotyronine. As with thyroid hormones leptin treatment resulted in partial restoration of testosterone and luteinizing hormone (fig. 1). Figure 2 illustrates the time course of leptin action on corticosterone and thyroxine during fasting. A single leptin injection at the end of a 48-hour fast blunted the fall in thyroxine within 3 h and decreased corticosterone further after 6 h.

Since nutritional status is an important determinant of reproduction including the estrus cycle, we determined whether leptin administration could prevent the delay of cyclicity brought on by fasting [56]. Leptin treatment restored normal estrus cycles in 70% of fasted mice [16]. There was a corresponding increase in estradiol levels in response to leptin administration (fig. 1c). As with rodents, there is an association between hypoleptinemia and disruption of reproductive function in humans. Laughlin and Yen [57] observed that female athletes with decreased mean leptin levels but intact diurnal leptin rhythm had normal menstrual cycles. In contrast, blunting of the nocturnal leptin surge was associated with amenorrhea. Recent studies have also shown that leptin deficiency or insensitivity from mutations of the *obese* and leptin receptor genes in humans leads to central hypogonadism and failure of pubertal development [38, 39].

The effect of leptin on thryoid function during starvation has been studied further in rats [17]. As observed in mice, leptin prevented the fall in triiodotyronine and thyroxine [16, 17]. Fasting for up to 64 h did not alter the level of thyroid-stimulating hormone (TSH) or fraction of thyroid hormone bound to protein [17]. In parallel with the fall in thyroid hormone, prothyrotropin-releasing hormone (proTRH) mRNA was decreased in the hypothalamic paraventricular nucleus but not other brain regions. Leptin

Table 1. Effect of fasting and leptin on metabolic parameters and hormones

	Fed	Fast 24 h	Fast 24 h + L	Fast 48 h	Fast 48 h + L
Body weight, % initial	99.3 ± 1.2	92.1 ± 2.4	92.4 ± 3.1	85.2 ± 3.2*	84.8 ± 2.8*
Glucose, mg/dl	148 ± 4.5	72 ± 1.5*	69 ± 3.5*	66 ± 2.1*	63 ± 2.4*
Beta-hydroxybutyrate, mg/dl	2.85 ± 0.35	ND	ND	19.3 ± 2.52*	17.2 ± 1.54*
BUN, mg/dl	7.8 ± 1.0	ND	ND	6.3 ± 0.6	6.8 ± 0.6
Insulin, ng/ml	1.32 ± 0.34	0.24 ± 0.04*	0.2 ± 0.04*	0.22 ± 0.05*	0.21 ± 0.02*
Leptin, ng/ml	4.2 ± 0.5	2.0 ± 0.3*	4.1 ± 0.7^{+}	1.4 ± 0.1*	3.8 ± 0.5^{+}
Cort, ng/ml	144 ± 31	575 ± 48*	470 ± 39*	315 ± 32*	218 ± 19$^{*,+}$
ACTH, pg/ml	57 ± 6.5	114 ± 16*	87 ± 13*	94 ± 5.4*	48 ± 4.7$^{*,+}$
Thyroxine, μg/dl	4.5 ± 0.2	2.6 ± 0.5*	3.3 ± 0.4*	1.5 ± 0.2*	2.3 ± 0.1$^{*,+}$
T3, ng/dl	114 ± 5.0	ND	ND	50 ± 5.0*	65 ± 2.8$^{*,+}$

Male C57B1/6 mice were housed in plastic cages under constant environmental conditions, 12-hour dark (18.00–06.00 h) and 12 hour light (06.00–18.00 h) cycles and allowed ad libitum access to normal chow and water. At age 7–10 weeks they were divided into 5 groups. One group continued to feed ad libitum, while the others were fasted for 24 or 48 h and received twice daily intraperitoneal injections of either murine leptin (1 μg/g body weight in 100 μl saline vehicle) or vehicle alone. All mice had continuous access to water. Leptin was administered at 09.00 and 18.00 h, respectively, and the mice were killed by decapitation 12 h after the last injection. Truncal blood was obtained for measurement of glucose, betahydroxybutyrate and urea nitrogen by enzyme assay and hormone measurement by radioimmunoassay [16]. Data are means ± SEM; n = 10–13 per group. *$p < 0.05$ compared with fed mice; $^{+}p < 0.05$ compared with fasted mice by ANOVA and Fisher protected least significant difference test. ACTH = Adrenocorticotropic hormone; BUN = blood urea nitrogen; T3 = triiodothyronine; ND = not measured.

prevented the fall in proTRH mRNA [17]. The possibility that low leptin levels lead to suppression of the hypothalamic-pituitary-thyroid axis is supported by recent findings in humans with leptin receptor mutation [39]. Leptin-insensitive humans have low free thyroxine levels, normal TSH and an exaggerated response to TRH stimulation, features consistent with central hypothyroidism [39].

A potential role of leptin in the regulation of the somatotropic axis has been demonstrated in rats [18, 58]. Carro et al. [18] have shown that spontaneous growth hormone secretion in rats can be blocked by immunoneutralization of leptin. Intracerebroventricular leptin administration did not affect growth hormone secretion in fed rats but prevented the starvation-induced fall in growth hormone [18]. The latter observation suggests a central action of leptin on the somatotropic axis, and is supported by the finding that leptin inhibits

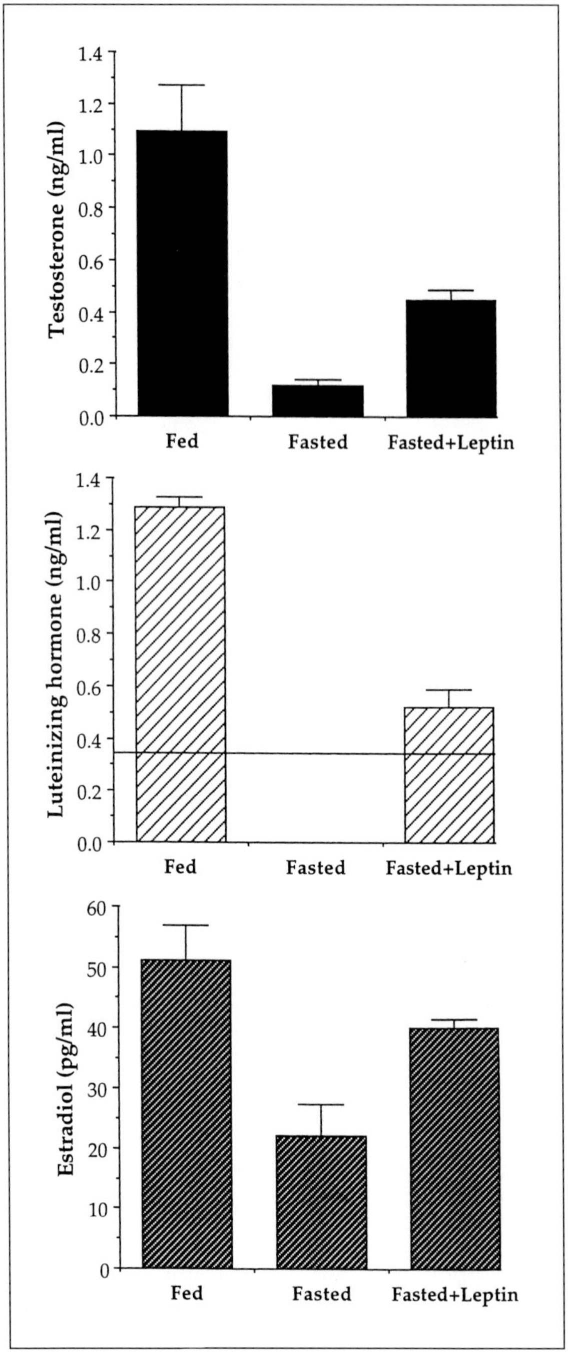
Testosterone (ng/ml)
1.4
1.2
1.0
0.8
0.6
0.4
0.2
0.0
Fed
Fasted
Fasted+Leptin
Luteinizing hormone (ng/ml)
1.4
1.2
1.0
0.8
0.6
0.4
0.2
0.0
Fed
Fasted
Fasted+Leptin
Estradiol (pg/ml)
60
50
40
30
20
10
0
Fed
Fasted
Fasted+Leptin

1a

1b

1c

somatostatin release as well as somatostatin mRNA levels [58]. Leptin may also regulate growth hormone levels in humans, as has been demonstrated by impaired growth hormone secretion and growth retardation in humans with leptin receptor mutations [39].

Although leptin regulates hormone secretion from the pituitary, adrenal glands and gonads [59–61], it is more likely to influence the diverse effects of starvation on neuroendocrine function through direct action in the brain. The ability of leptin to restore proTRH mRNA in starved rats, as well as regulate corticotropin-releasing hormone and gonadotropin-releasing hormone levels is consistent with this view [17, 59, 62, 64]. NPY has been suggested as a potential mediator of the effects of leptin on food intake and neuroendocrine function [16, 47, 53]. Hypothalamic NPY levels are elevated in states of leptin deficiency including starvation [16, 17] and decreased by leptin treatment [16, 53]. Previous studies have also shown that NPY can activate the pituitary-adrenal axis and suppress the growth hormone and reproductive axes [65–67]. However, since NPY-deficient mice have a similar neuroendocrine response to starvation as normal mice [68] and are capable of responding to leptin treatment [69], it is more likely that other hypothalamic factors mediate the central actions of leptin during starvation.

The possibility that leptin may cooperate with other signals to influence the adaptation to fasting is suggested by our recent findings in mice subjected to a classic food restriction paradigm [70]. Diurnal levels of leptin and other metabolic hormones such as glucocorticoids, insulin and thyroxine appeared to be entrained by the timing of feeding [70]. In ad-libitum-fed mice, the nadir of leptin preceded the surge in corticosterone and onset of maximal feeding during the dark cycle. Insulin levels rose during the dark cycle while thyroxine increased in the light cycle. Restriction of ad libitum feeding to 4 h of the light cycle led to an initial decrease in body weight followed by increased food intake and normalization of body weight after 1 week. After 2 weeks of restricted feeding there was a parallel shift of peak levels of leptin, glucocorticoids, insulin and thyroxine. In agreement with previous studies, corticosterone

Fig. 1. a, b Effect of 48 h of fasting and leptin administration on male sex hormones. Male C57B1/6 mice housed and treated as in table 1 were used. Plasma testosterone (*a*) and LH (*b*) were measured by radioimmunoassay [16]. Data are means ± SEM, n = 10–13 per group. The limit of detection of LH (0.36 ng/ml) is indicated in figure 1*b*. LH was not detectable in fasted mice which were injected with saline vehicle alone. *c* Effect of 48 h of fasting and leptin on estradiol levels. Female C57B1/6 mice were housed singly under ambient conditions as in table 1 and allowed free access to food and water. Daily vaginal smears were performed from age 8 weeks for estrus 3 cycles. Mice with regular 3 μg/g body weight or saline vehicle. They were killed 12 h later and serum estradiol was measured by radioimmunoassay (ICN, Costa Mesa, Calif., USA).

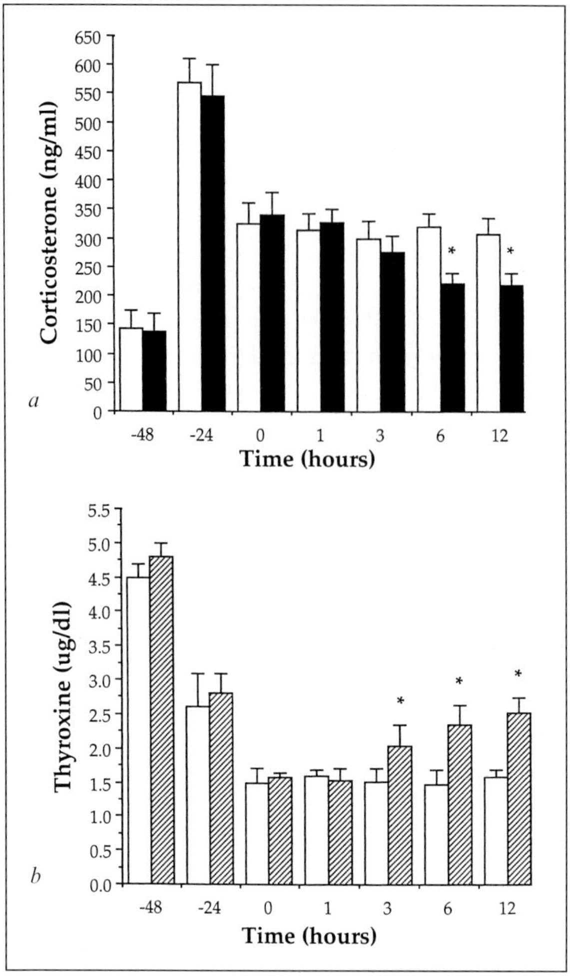

Fig. 2. Time course of leptin action on corticosterone (*a*) and thyroxine (*b*) after leptin administration to fasted mice. Male C57BL/6 mice housed as in table 1 were fasted for 48 h and injected intraperitoneally with a single dose of either murine leptin (2 μg/g body weight in 100 μl saline) or vehicle. Groups of mice (n = 5) were killed by decapitation immediately after the injection (time 0; 09.00 h) and 1, 3, 6, and 12 h later. Serum corticosterone and thyroxine were measured by radioimmunoassay (ICN, Costa Mesa, CA). Data are means ± SEM. Dark bars = Leptin treated; open bars = saline treated. *$p < 0.05$ compared with saline-treated mice by ANOVA and Fisher protected least significant difference test.

levels peaked during the light cycle before the onset of feeding and declined shortly after feeding [70–72]. The reciprocal relation between corticosterone and leptin was maintained such that the nadir of leptin occurred before feeding, while the rise in leptin coincided with the fall in corticosterone after feeding [71].

The fall in leptin with starvation is rapid and out of proportion to fat mass. For example, serum leptin falls by 50% while fat mass decreases by only 0.5% after 24 h of fasting in humans [42]. Similarly, fasting decreases leptin mRNA by 50–80% during short-term fasting in rodents [15, 73, 74]. Since insulin levels decrease in parallel with leptin during fasting and insulin regulates leptin synthesis [13, 15, 16, 73–75], it is possible that the rapid fall in leptin in the early phase of starvation is regulated in part by the fall in insulin. In agreement with this view, prevention of the fall in insulin during fasting by glucose infusion also prevents leptin levels from falling, and elevation of insulin leads to increased leptin synthesis [15, 73–75].

There is considerable overlap in the distribution of hypothalamic targets of leptin, insulin and glucocorticoids. It is therefore possible that the adaptation to starvation involves a coordinated action of these hormone signals on similar neural circuits in the brain. Activation of the hypothalamic-pituitary-adrenal axis by leptin may prolong survival by increasing food-seeking behavior [5], in addition to stimulating gluconeogenesis to provide glucose for use by vital organs such as the brain [1, 2, 5]. Leptin-mediated suppression of the thyroid, growth hormone and reproductive axes is likely to influence survival by decreasing thyroid thermogenesis and the high energy cost of growth and reproduction.

In contrast to its ability to influence hormone levels during starvation, the role of leptin in fuel homeostasis is yet to be determined. Leptin did not affect metabolic parameters such as body weight, blood glucose, beta-hydroxybutyrate, plasma insulin and blood urea nitrogen during starvation [16, 17] (table 1). Moreover, leptin had little effect on food intake and weight gain after starvation [16]. The lack of a metabolic response to leptin may be due to the dose of leptin used in the above experiments [16, 17] (table 1). However, it is also possible that the neuroendocrine axis is more sensitive to falling leptin levels during starvation. One can envisage a model for the adaptation to starvation in which the fall in insulin mediates peripheral and perhaps central responses to maintain fuel homeostasis, while the fall in leptin activates the adrenal axis and suppresses sex, growth and thyroid hormones. The existence of such a system for the adaptation to starvation would of course have to be proven.

Unlike adults, leptin does not appear to be regulated by adipose tissue mass or by short-term food deprivation during the early postnatal period

[70, 76]. A rise in postnatal leptin precedes the onset of puberty in humans and rodents [70, 76] and establishment of adult patterns of glucocortiocoids and thyroxine in mice [70]. Fasting for up to 12 h does not alter leptin and glucocorticoid levels in mouse pups [70]. Possible explanations of the lack of response of leptin and glucocorticoids to fasting include immaturity of the neuroendocrine axis and differences in the kinetics of these hormones during the postnatal period. The functional implications of the age difference in leptin regulation awaits further study.

Conclusions

The discovery of the adipose-derived hormone leptin has generated enormous interest in the molecular interaction between nutrition and effector circuits in the brain. I have reviewed evidence on the potential role of leptin as a mediator of diverse endocrine responses during starvation. Leptin is well positioned to be a sensor of energy stores as well as a signal to inform the brain about the decline in energy stores. The adaptation to starvation mediated by leptin, insulin, glucocorticoids and other signals is likely to promote survival by facilitating efficient use of fuel substrates as well as stimulating autonomic, endocrine and behavioral responses to increase food intake and conserve energy.

Acknowledgments

This work was supported by NIH DKR3728082 and a grant from Eli Lilly and Co. to J.S. Flier. R.S.A. was a Pfizer Fellow.

References

1 Cahill JG, Herrera MG, Morgan AP, Soeldner JS, Steinke J, Levy PL, Reichard GA, Kipnis DM: Hormone-fuel interrelationships during fasting. J Clin Invest 1966;45:1751–1769.
2 Sims EA, Horton ES: Endocrine and metabolic adaptation to obesity and starvation. Am J Clin Nutr 1968;21:1455–1470.
3 Sakaguchi T, Arase K, Fisler JS, Bray GA: Effect of starvation and food intake on sympathetic activity. Am J Physiol 1988;255:R284–R288.
4 Bergendahl M, Vance ML, Iranmanesh A, Thorner MO, Veldhuis JD: Fasting as a metabolic stress paradigm selectively amplifies cortisol secretory burst mass and delays the time of maximal nyctohemeral cortisol concentrations in healthy men. J Clin Endocrinol Metab 1996;81:692–699.
5 Dallman MF, Strack AM, Akana SF, Bradbury MJ, Hanson EF, Scribner KA, Smith M: Feast and famine: Critical role of glucocorticoids with insulin in daily energy flow. Front Neuroendocrinol 1993;14:303–347.

6 Pirke K, Schweiger U, Strowitzki T, Tuschl RJ, Laesck RG, Broocks A, Huber B, Middendorf R: Dieting causes menstrual irregularities in normal weight young women through impairment of episodic luteinizing hormone secretion. Fertil Steril 1989;51:263–268.
7 Campbell GA, Kurcz M, Marshall S, Meitex J: Effects of starvation in rats on serum levels of follicle stimulating hormone, luteinizing hormone, thyrotropin, growth hormone and prolactin: Response to LH-releasing hormone and thyrotropin-releasing hormone. Endocrinology 1977;100: 580–587.
8 Harris RC, Fang S-L, Azizi F, Lipworth L, Vagenakis AG, Braverman LE: Effect of starvation on hypothalamic-pituitary-thyroid function in the rat. Metabolism 1978;27:1074–1083.
9 Tannenbaum GS, Rostad O, Brazeau P: Effects of prolonged food deprivation on the ultradian growth hormone rhythm and immunoreactive somatostatin tissue levels in the rat. Endocrinology 1979;104:1733–1738.
10 Merimee TJ, Zapf J, Froesch ER: Insulin-like growth factors in the fed and fasted states. J Clin Endocrinol Metabl 1982;55:999–1002.
11 Maffei MJ, Halaas J, Rayussin E, Pratley RE, Lee GM, Zhang Y, Fei H, Kim S, Lallone R, Ranganathan S: Leptin levels in human and rodent: Measurement of plasma leptin and obmRNA in obese and weight-reduced subjects. Nat Med 1995;1:1155–1161.
12 Frederich RC, Hamann A, Anderson S, Lollman B, Lowell BB, Flier JS: Leptin levels reflect body lipid content in mice: Evidence for diet-induced resistance to leptin action. Nat Med 1995;1: 1311–1314.
13 Considine RV, Sinha MK, Heiman ML, Kriauciunas A, Stephens TW, Nyce MR, Ohannesian JP, Marco CC, McKee LJ, Bauer TL, Caro JF: Serum immunoreactive leptin concentrations in normal weight and obese humans. N Engl J Med 1996;334:292–295.
14 Frederich RC, Lollman B, Hamann A, Napolitano-Rosen A, Kahn B, Lowell BB, Flier JS: Expression of ob mRNA and its encoded protein in rodents. J Clin Invest 1995;96:1658–1663.
15 Saladin R, Devos P, Guerre-Millo M, Leturge A, Girard J, Staels B, Auwerx J: Transient increase in obese gene expression after food intake or insulin administration. Nature 1995;377:527–529.
16 Ahima RS, Prabakaran D, Mantzoros C, Qu D, Lowell B, Maratos-Flier E, Flier JS: Role of leptin in the neuroendocrine response to fasting. Nature 1996;382:250–252.
17 Legradi G, Emerson CH, Ahima RS, Flier JS, Lechan RM: Leptin prevents fasting-induced suppression of prothyrotropin-releasing hormone messenger ribonucleic acid in neurons of the hypothalamic paraventricular nucleus. Endocrinology 1997;138:2569–2576.
18 Carro E, Senaris R, Considine RV, Casanueva FF, Dieguez C: Regulation of in vivo growth hormone secretion by leptin. Endocrinology 1997;138:2203–2206.
19 Kennedy GC: The role of depot fat in the hypothalamic control of food intake in the rat. Proc R Soc Lond (Biol) 1953;140:578–596.
20 Harris RBS, Kasser TR, Martin RJ: Dynamics of recovery of body composition after overfeeding, food restriction or starvation of mature female rats. J Nutr 1986;116:2536–2546.
21 Seeley RJ, Matson CA, Chavez M, Woods SC, Dallman MF, Schwartz MW: Behavioral, endocrine, and hypothalamic responses to involuntary overfeeding. Am J Physiol 1996;271:R819–R823.
22 Coleman DL: Effects of parabiosis of obese with diabetes and normal mice. Diabetologia 1973;9: 294–298.
23 Zhang Y, Proenca R, Maffei M, Barone M, Leopold L, Friedman JM: Positional cloning of the mouse obese gene and its human homologue. Nature 1994;372:425–432. Erratum, Nature 1995; 374:479].
24 Campfield LA, Smith FJ, Guisez J, Devos R, Burn P: Recombinant mouse OB protein: Evidence for a peripheral signal linking adiposity and central neural networks. Science 1995;269:546–549.
25 Pelleymounter M, Cullen M, Baker M, Hecht R, Winters D, Boone T, Collins F: Effects of the obese gene product on body weight regulation in ob/ob mice. Science 1995;269:540–543.
26 Halaas J, Gajiwala K, Maffei M, Cohen S, Chait B, Rabinowitz D, Lallone R, Burley S, Friedman J: Weight-reducing effects of the plasma protein encoded by the obese gene. Science 1995;269: 543–546.

27 Tartaglia L, Dembski M, Weng X, Deng N, Culpepper J, Devos R, Richards G, Campfield L, Clark F, Deeds J, Muir C, Sanker S, Moriarty A, Moore K, Smutko J, Mays G, Woolf E, Monroe C, Tepper R: Identification and expression cloning of a leptin receptor. Cell 1995;83:1263–1271.

28 Lee GH, Proenca R, Montez JM, Carroll K, Darvischzadeh J, Lee J, Friedman J: Abnormal splicing of the leptin receptor in diabetic mice. Nature 1996;379:632–635.

29 Chua SC, Chung WK, Wu-Peng XS, Zhang Y, Liu S-M, Tartaglia L, Leibel RL: Phenotypes of mouse *diabetes* and rat *fatty* due to mutations in the OB (leptin) receptor. Science 1996;271:994–996.

30 Bauman H, Morella KK, White DW, Dembski M, Bailon PS, Kim H, Lai C-F, Tartaglia LA: The full length leptin receptor has signaling capabilities of interleukin 6-type cytokine receptors. Proc Natl Acad Sci USA 1996;93:8374–8378.

31 Fei H, Okana HJ, Li C, Lee G-H, Zhao C, Darnell R, Friedman JM: Anatomic localization of aternatively spliced leptin receptors (Ob-R) in mouse brain and other tissues. Proc Natl Acad Sci USA 1997;94:7001–7005.

32 Mercer JG, Hoggard N, Williams LM, Lawrence CB, Hannah LT, Trayhurn P: Localization of leptin mRNA and the long form splice variant (Ob-Rb) in mouse hypothalamus and adjacent brain regions by in situ hybridization. FEBS Lett 1996;387:113–116.

33 Elmquist JK, Ahima RS, Maratos-Flier E, Flier JS, Saper CB: Leptin activates neurons in the ventrobasal hypothalamus and brainstem. Endocrinology 1997;138:839–842.

34 Elmquist JK, Ahima RS, Elias CS, Flier JS, Saper CB: Leptin activates distinct projections from the dorsomedial and ventromedial hypothalamic nuclei. Proc Natl Acad Sci USA 1998;95:741–746.

35 Banks WA, Kastin AJ, Huang W, Jaspan JB, Maness LM: Leptin enters the brain by a saturable system independent of insulin. Peptides 1996;17:305–311.

36 Elmquist JK, Bjorbaek C, Ahima RS, Flier JS, Saper CB: Distributions of leptin receptor mRNA isoforms in the rat brain. J Comp Neurol 1998;395:535–547.

37 Montague CT, Farooqui S, Whitehead JP, Soos MA, Rau H, Wareham NJ, Sewter CP, Digby JE, Mohammed SN, Hurst JA, Cheetham CH, Earley AR, Barnett AH, Prins JB, O'Rahilly S: Congenital leptin deficiency is associated with severe early onset obesity in humans. Nature 1997;387:903–908.

38 Strobel A, Issad T, Camoin L, Ozata M, Strosberg AD: A leptin missence mutation associated with hypogonadism and morbid obesity. Nat Genet 1998;18:213–215.

39 Clement K, Vaisse C, Lahlous N, Cabroll S, Pelloux V, Cassuto D, Gourmelen M, Dina G, Chambaz J, Lacorte J-M, Basdervant A, Bougnerest P, Lebouc Y, Frouguel P, Guy-Grand B: A mutation in the human leptin receptor gene causes obesity and pituitary dysfunction. Nature 1998;392:398–401.

40 Schwartz MW, Figlewicz DP, Baskin DG, Woods SC, Porte D Jr: Insulin in the brain: A hormonal regulator of energy balance. Endocr Rev 1992;13:387–414.

41 Strack AM, Sebastian RJ, Schwartz MW, Dallman MF: Glucocorticoids and insulin: Reciprocal signals for energy balance. Am J Physiol 1995;268:R142–R149.

42 Boden G, Chen X, Mozzoli M, Ryan I: Effect of fasting on serum leptin in normal human subjects. J Clin Endocrinol Metab 1996;81:3419–3423.

43 Sinha MK, Ohannesian JP, Heiman ML, Kriaucinnas A, Stephens MW, Magosin S, Marco C, Caro JF: Nocturnal rise of leptin in lean, obese and non-insulin dependent diabetes mellitus subjects. J Clin Invest 1996;97:1344–1347.

44 Dawson R, Pelleymounter MA, Millard W, Liu S, Eppler B: Attenuation of leptin-mediated effects by monosodium glutamate-induced arcuate nucleus damage. Am J Physiol 1997;273:E202–E206.

45 Hakansson M-L, Brown H, Ghilardi N, Skoda RC, Meister B: Leptin receptor immunoreactivity in chemically defined target neurons in the hypothalamus. J Neurosci 1998;18:559–572.

46 Mercer JG, Hoggard N, Williams LM, Lawrence CB, Hannah LT, Morgan PJ, Trayhurn P: Coexpression of leptin receptor and preproneuropeptide Y mRNA in arcuate nucleus of mouse hypothalamus. J Neuroendocrinol 1996;8:733–735.

47 Schwartz M, Seeley RJ, Campfield LA, Burn P, Baskin DG: Identification of targets of leptin action in rat hypothalamus. J Clin Invest 1996;98:1101–1106.

48 Thornton JE, Cheung CC, Clifton DK, Steiner RA: Regulation of hypothalamic proopiomelanocortin mRNA by leptin in mice. Endocrinology 1997;138:5063–5066.
49 Bray GA, York DA: Hypothalamic and genetic obesity in experimental animals: An autonomic and endocrine hypothesis. Physiol Rev 1979;59:719–790.
50 Garthwaite TL, Martinson DR, Tseng LF, Hagen TC, Menahan LA: A longitudinal hormonal profile of the genetically obese mouse. Endocrinology 1980;107:671–676.
51 Mcginnis R, Walker J, Margules D, Aird F, Redei E: Dysregulation of the hypothalamus-pituitary-adrenal axis in male and female genetically obese (ob/ob) mice. J Neuroendocrinol 1992;4:765–771.
52 Brady LS, Smith MA, Gold PW, Herkenham M: Altered expression of hypothalamic neuropeptide mRNAs in food restricted and food-deprived rats. Neuroendocrinology 1990;52:441–447.
53 Stephens TW, Basinski M, Bristow PK, Bue-Valleskey JM, Burgett SG, Craft L, Hale J, Hoffman J, Hsiung HM, Kriauciunas A, MacKeller W, Rostock PR Jr, Schoner B, Smith D, Tinsley PC, Zhang X-Y, Heiman M: The role of neuropeptide Y in the antiobesity action of the obese gene product. Nature 1995;377:530–532.
54 Chehab F, Lim M, Lu R: Correction of the sterility defect in homozygous obese female mice by treatment with the human recombinant leptin. Nat Genet 1996;12:318–320.
55 Barash IA, Cheung CC, Weigle DS, Hongping R, Kagigting EB, Kuijper JL, Clifton DL, Steiner RA: Leptin is a metabolic signal to the reproductive system. Endocrinology 1997;137:3144–3147.
56 Bronson FH: Effect of food manipulation on the GnRH-LH-estradiol axis in young female rats. Am J Physiol 1988;254:R616–R621.
57 Laughlin GA, Yen SS: Hypoleptinemia in women athletes: Absence of a diurnal rhythm with amenorrhea. J Clin Endocrinol Metab 1997;82:318–321.
58 Quinetela M, Senaris R, Heiman ML, Casanueva FF, Dieguez C: Leptin inhibits in vitro hypothalamic somatostatin secretion and somatostatin mRNA levels. Endocrinology 1997;138:5641–5644.
59 Yu WH, Kimura M, Walczewska A, Karanth S, McCann SM: Role of leptin in hypothalamic-pituitary function. Proc Natl Acad Sci USA 1997;94:1023–1028.
60 Bornstein SR, Uhlmann K, Haidan A, Ehrhart-Bornstein M, Scherbaum WA: Evidence for a novel peripheral action of leptin as a metabolic signal to the adrenal gland: Leptin inhibits cortisol release directly. Diabetes 1997;46:1235–1238.
61 Zachow RJ, Magoffin DA: Direct intraovarian effects of leptin: Impairment of the synergistic action on insulin-like growth factor I on follicle stimulating hormone dependent estradiol 17 beta production by rat ovarian granulosa cells. Endocrinology 1997;138:847–850.
62 Costa A, Poma A, Martignoni E, Nappi G, Ur E, Grossman A: Stimulation of corticotropin-releasing hormone release by obese (ob) gene product, leptin, from hypothalamic explants. Neuroreport 1997; 8:1131–1134.
63 Raber J, Chen S, Mucke L, Feng L: Corticotropin-releasing factor and adrenocorticotropic hormone as potential central mediators of OB effects. J Biol Chem 1997;272:15057–15060.
64 Heiman ML, Ahima RS, Craft LS, Schoner B, Stephens TW, Flier JS: Leptin inhibition of the hypothalamic-pituitary-adrenal axis in response to stress. Endocrinology 1997;138:3859–3863.
65 White JD: Neuropeptide Y: A central regulator of energy homeostasis. Regul Peptides 1993;49: 93–107.
66 Wahlsedt C, Skagerberg G, Ekman R, Helig M, Sundler F, Hakanson R: Neuropeptide Y (NPY) in the area of the hypothalamic paraventricular nucleus activates the pituitary-adrenocortical axis in the rat. Brain Res 1987;417:33–38.
67 Catzeflis C, Pierroz DD, Rohner-Jeanrennaud F, Rivier JE, Sizonenko PC, Aubert ML: Neuropeptide Y administered chronically into the lateral ventricle profoundly inhibits both the gonadotropic and the somatotropic axis in intact adult female rats. Endocrinology 1993;132:224–234.
68 Erickson JC, Ahima RS, Hollopeter G, Flier JS, Palmiter RD: Endorine function of neuropeptide Y knockout mice. Regul Pept 1997;70:199–202.
69 Erickson JC, Clegg KE, Palmiter RD: Sensitivity to leptin and susceptibility to seizures of mice lacking neuropeptide Y. Nature 1996;381:415–418.
70 Ahima RS, Prabakaran D, Flier JS: Postnatal leptin surge and regulation of circadian rhythm of leptin by feeding: Implications for energy homeostasis and neuroendocrine function. J Clin Invest 1998;101:1020–1027.

71 Itoh S, Katsuura G, Hirota R: Conditioned circadian rhythm of plasma corticosterone in the rat induced by food restriction. Jpn J Physiol 1980;30:365–375.
72 Morimoto Y, Arisue K, Yamamura Y: Relationship between circadian rhythm of food intake and that of plasma corticosterone and effect of food restriction on circadian adrenocortical rhythm. Neuroendocrinology 1977;23:212–222.
73 Trayhurn P, Thomas MEA, Duncan JS, Rayner DV: Effects of fasting and refeeding on ob gene expression in white adipose tissue of lean and obese mice. FEBS Lett 1995;368:488–490.
74 MaDougald OA, Hwang CS, Fan H, Lane MD: Regulated expression of the obese gene product (leptin) in white adipose tissue and 3T3-L1 adipocytes. Proc Natl Acad Sci USA 1995;92:9034–9037.
75 Cusin I, Sainbury A, Doyle P, Rohner-Jeanrenaud F, Jeanrenaud B: The obse gene and insulin: A relationship leading to clues to the understanding of obesity. Diabetes 1995;44:1467–1479.
76 Mantzoros C, Flier JS, Rogol AD: A longitudinal assessment of hormonal and physical alterations during normal puberty in boys. V. Rising leptin levels may signal the onset of puberty. J Clin Endocrinol Metab 1997;82:1066–1070.

Rexford S. Ahima, MD, PhD, Division of Endocrinology, Diabetes and Metabolism,
Department of Medicine, University of Pennsylvania Medical Center,
611 Clinical Research Bldg, 415 Curie Blvd, Philadeplphia, PA 19104 (USA)
Tel. +1 215 898 0210, Fax +1 215 898 5408, E-Mail r_ahima@hotmail.com

Ur E (ed): Neuroendocrinology of Leptin.
Front Horm Res. Basel, Karger, 2000, vol 26, pp 57–70

Leptin Regulation of Proopiomelanocortin

Charles Mobbs, Tooru Mizuno

Mt. Sinai School of Medicine, Fishberg Center for Neurobiology, New York, N.Y., USA

Agouti, the First Cloned Obese Gene, Antagonizes α-MSH

The first obese gene (that is, a gene a mutation in which can cause obesity) to be cloned was the *agouti* gene [1]. Mutations in the *agouti* locus, especially the lethal yellow (A^y) mutation, promote the development of an *agouti* coat color on an otherwise black background (hence its name), but of much more general interest the A^y allele also promotes obesity, insulin resistance, and a variety of neoplasms. It is important to recognize that the A^y is a lethal recessive, so that only heterozygotes expressing this allele are viable and can be studied. Thus, although heterozygous mice expressing the A^y allele do not exhibit as extreme a phenotype as homozygous *ob/ob* (leptin-deficient) mice, the obesity of heterozygous A^y/+ mice is roughly as profound, if not more profound, as the obesity of heterozygous ob/+ mice. Because of its association with hair color, the *agouti* locus has long been the subject of genetic analysis, which, along with the use of a radiation-induced inversion mutation [2], facilitated the cloning of the *agouti* gene and its product. Upon cloning it was immediately evident that the nature of the mutation did not entail inactivation of the normal protein product through a sense or nonsense mutation, but rather involved a mutation in the promoter region of the gene, leading to overexpression of the gene in tissues which did not normally express the protein [1, 3].

Biochemical studies had indicated that the product of the *agouti* locus acts by stimulating the conversion of dopaquinone to phaeomelanin in melanocytes, which has the effect of lightening coat color [4]. Alternatively, coat color darkens due to conversion of dopaquinone to eumelanin, a process which in turn is stimulated by α-melanocyte-stimulating hormone (α-MSH) [5], a hormone which, as its name suggests, was originally isolated on the basis of its ability to stimulate melanocytes. Genetic and biochemical analysis had suggested that the product of the *agouti* locus acts through another gene defined by the *extension* locus, since mice homozygous for the e/e allele are similarly agouti colored, leading to the hypothesis that the *extension* locus codes for a receptor through which the *agouti* protein (hypothetically a secreted peptide) acts [4]. Injection of α-MSH could stimulate darkening in (and thus overcome the effect of) *agouti* heterozygous A^y/– mice, but α-MSH had no effect in *agouti* homozygous e/e mice [6]. These and other studies suggested that the protein product of the *agouti* gene interfered with the action of α-MSH, which normally acted through the wild-type extension locus [7]. This hypothesis was supported by the cloning of the *agouti* gene protein, which contained a signal peptide suggesting that the protein was secreted [1]. Together with the observation that the A^y allele involved ectopic expression of the *agouti* protein, this raised the possibility that the obesity associated with the A^y locus was due to ectopic expression of an antagonist to α-MSH [1]. This hypothesis thus implied that impairment in sensitivity to α-MSH tone can lead to obesity. However, there was little other data at that time which would support the hypothesis that α-MSH plays a major role in regulating body weight.

Inactivation of α-MSH Receptor Produces Obesity

At about the same time the *agouti* gene was cloned, a family of receptors for α-MSH were also cloned [8]. The availability of both receptors and ligands made it possible to directly test the hypotheses which had previously been inferred indirectly on the basis of genetic and biochemical evidence. For example, the cloning of members of this family quickly led to the demonstration that the *extension* locus does indeed encode a receptor for α-MSH [9], as predicted. Furthermore, partially purified *agouti* protein acted as a high-affinity competitive agonist for at least two α-MSH receptor subtypes, MC1-R and MC4-R [10]. This latter observation was particularly pertinent because unlike other members of the α-MSH receptor family, the MC4-R receptor (along with the MC3-R receptor) appeared to be expressed primarily in brain [11].

Although suggestive, these observations still did not directly prove that the antagonism of the α-MSH receptor was the specific property of the *agouti* gene product which caused obesity. The definitive study demonstrating the role of the α-MSH receptor system in the control of body weight involved the targeted disruption of the MC4-R receptor, which lead to an obese phenotype strikingly similar to that of *agouti* mice, including hyperphagia, increased adiposity, and increased linear growth [12]. At about the same time, specific nonpeptide agonists and antagonists were being developed which allowed the pharmacological assessment of the role of MC3-R and MC4-R in feeding

[13]. Intracerebroventricular infusion of the agonist decreased feeding in four models of hyperphagia, and administration of the agonist significantly enhanced nocturnal feeding [13]. Taken together, these studies strongly supported the hypothesis that α-MSH, acting through MC4-R receptors, exerts a tonic inhibition of feeding and probably other anabolic processes, and that impairment in the tone of α-MSH can lead to obesity.

Fasting Reduces Hypothalamic Expression of POMC, which Codes for the Precursor of α-MSH

It has long been appreciated that endogenous opioids, and in particular β-endorphin, generally stimulate feeding [14, 15]. These observations led to the concept that nutritional status might therefore regulate expression in the brain of the genes coding the endogenous opioids. β-endorphin is coded by the proopiomelanocortin (POMC) gene, which was the first opioid gene cloned [16]. When it became clear that the POMC gene exhibited a highly restricted expression in the brain, limited almost exclusively to the hypothalamus [17], it became of interest to assess if the expression of hypothalamic POMC was regulated by nutritional status. Therefore, Brady et al. [18] examined the effect of a 72-hour fast as well as caloric restriction for 2 weeks and found that these treatments led to a decrease in the levels of hypothalamic POMC mRNA. This result was corroborated by a subsequent report [19]. A much more detailed study demonstrated that hypothalamic POMC mRNA, as well as mRNAs for other opioid genes (prodynorphin and proenkephalin) decreased in proportion to the degree of food deprivation and body weight loss, although POMC mRNA was the most sensitive to caloric restriction [20]. Since endogenous opioids stimulate feeding, and body weight loss is usually accompanied by a compensatory hyperphagia, it might have been expected that fasting would lead to an increase, rather than a decrease, in hypothalamic POMC mRNA. For example, like endogenous opioids, hypothalamic neuropeptide Y (NPY) increases feeding, and the expression of NPY mRNA increases as body weight decreases [21, 22].

However, the apparently anomalous regulation of POMC mRNA becomes more explicable upon consideration that POMC codes not just for β-endorphin, but also for α-MSH. As indicated above, considerable evidence had indicated that α-MSH decreases feeding behavior, in opposition to the effects of β-endorphin. This activity of α-MSH was first suggested by the observation that infusion of ACTH (of which α-MSH constitutes the N-terminal portion) into the lateral ventricle inhibits feeding in rats [23]. A subsequent study by the same group demonstrated that both ACTH and α-MSH block the stimulation of feeding by a kappa opioid agonist [24]. The antagonism between α-MSH and β-endorphin had been observed in a wide variety of responses [reviewed by Poggioli et al., 24]. The biological significance of this antagonistic effects of two peptides synthesized from the same precursor was suggested by a study by Tsujii and Bray [25]. These investigators noted that the N-acetylated form of MSH, α-MSH, strongly inhibited food intake when infused into the third ventricle, whereas the desacetylated form had no effect on food intake. In contrast, the acetylated form of β-endorphin had no effect on food intake, whereas the nonacetylated form stimulated feeding behavior [25]. Therefore, the net effect of the opposed action of β-endorphin vs. α-MSH may depend on the extent of acylation, with high levels of acylation (in which most MSH is found) favoring the satiety effects, and low levels of acylation favoring the orexigenic effects. Thus the inhibition of POMC mRNA by fasting is consistent with the hypothesis that this inhibition produces compensatory hyperphagia by reducing the synthesis of α-MSH, which under circumstances of high acylation would produce anorexic or catabolic effects which dominate the orexigenic or anabolic effects of β-endorphin.

The physiological significance of results derived from the *agouti* gene may be questioned since it is clear that the normal physiological function of this gene product is to regulate skin color. However, the discovery of the *agouti*-related transcript in the hypothalamus demonstrated the physiological significance of the pathway [26].

Leptin Resistance in *Agouti* Mice

About 2 years after the *agouti* gene was cloned, the *obese* gene was cloned using positional cloning [27]. The predicted product of this gene, referred to as leptin [28], was a peptide secreted from adipose tissue, consistent with predictions made by Coleman. It was quickly apparent that obese A^y mice exhibited elevated expression of the leptin gene [29]. This result suggested that *agouti* mice were resistant to the effects of leptin, and since *agouti* acts by blocking effects of α-MSH, this hypothesis suggested that leptin might also act, in part, by stimulating the production of α-MSH, and thus that leptin would stimulate hypothalamic POMC mRNA.

Leptin Reverses Fasting Phenotype

It was soon evident that leptin gene expression was reduced during fasting [29, 30]. The pertinence of this observation became apparent by the elegant demonstration that injection of leptin would reverse many of the neuroendocrine effects of fasting, including elevated glucocorticoids and decreased thyroid hormone [31]. For example, as indicated above, fasting stimulates hypothalamic NPY mRNA [21, 22]; conversely, leptin decreases hypothalamic NPY mRNA [32]. Since, as described above, fasting decreases hypothalamic POMC mRNA, these observations suggested that leptin would stimulate hypothalamic POMC mRNA.

Leptin Receptor Colocalizes with Hypothalmic POMC mRNA

The positional cloning of leptin was followed quickly by the cloning of the receptor for leptin, which mapped to the diabetes locus [33, 34]. Mapping studies by in situ hybridization led to the observation that, depending on the precise criteria, between 30 and 90% of POMC-producing neurons also express the leptin receptor [35]. These studies were consistent with the hypothesis, suggested by the above results, that leptin might regulate hypothalamic POMC mRNA.

Leptin Stimulates Hypothalamic POMC mRNA

The above observations led to the simultaneous demonstration by three groups that hypothalamic POMC mRNA is stimulated by leptin [36–38]. A pivotal prediction of the hypothesis that leptin stimulates hypothalamic POMC mRNA is that, if true, hypothalamic POMC mRNA levels would be lower in *ob/ob* and *db/db* mice, which lack active leptin and are resistant to the effects of leptin, respectively, than in wild-type mice, and all three papers addressed this prediction, with variations.

In the first published report, Thornton et al. [36] used in situ hybridization to examine POMC mRNA expression in three groups: wild-type mice, *ob/ob* injected i.p. with leptin (2 μg/g body weight/day over 6 days), and *ob/ob* injected with saline and pair-fed to match the food intake of the leptin-injected *ob/ob* mice. In this study, detectable POMC-producing neurons in the rostral arcuate nucleus were about 50% more numerous in both leptin-injected *ob/ob* mice and wild-type mice compared to saline-injected *ob/ob* mice. Similarly, throughout the arcuate nucleus, wild-type expressed between 50 and 100% more grains per cell (with the greater effect in the rostral sections) than saline-injected *ob/ob* mice, whereas leptin-injected *ob/ob* mice exhibited about the same number of grains/cell as wild-type mice.

In the second published report, Schwartz et al. [37] also used in situ hybridization to examine POMC mRNA levels in *ob/ob* mice, but in addition examined *db/db* and effects of leptin on POMC mRNA levels in fasting rats.

These investigators confined their analysis to the rostral arcuate nucleus, which in their hands exhibited the most robust responses to fasting. Corroborating Thornton et al. [36], Schwartz et al. [37] reported that in the rostral arcuate nucleus POMC mRNA levels (based on hybridization area times pixels density, roughly the total signal) were about 2-fold higher in wild-type than in *ob/ob* mice, and about 3-fold higher in wild-type than in *db/db* mice. These results are consistent with those of Thornton et a1. [36], since presumably they include both the increased number of observable cells, as well as the increased number of grains per cell. Furthermore, i.p. injection of leptin (150 μg/mouse/day over 5 days) into *ob/ob* mice stimulated hypothalamic POMC mRNA levels by about 70% compared to saline-injected and pair-fed *ob/ob* mice. However, leptin injection did not produce any effect on POMC mRNA in *db/db* mice. In the same study, POMC mRNA levels were about 2-fold higher in ad-libitum-fed mice than in 48-hour-fasted mice. Furthermore, intercerebral ventricular injection of leptin (3.5 μg) into fasting rats elevated POMC mRNA levels about 40%.

The third report, by Mizuno et al. [38] examined leptin, POMC, NPY, and MCH (melanin-concentrating hormone) mRNA levels in fasting, *ob/ob* and *db/db* mice, using Northern blot analysis for leptin, POMC, and NPY mRNAs, and using in situ hybridization for POMC, NPY, and MCH mRNAs. The effect of leptin replacement in male and female *ob/ob* mice on POMC and NPY mRNA levels was also assessed by Northern blot and in situ hybridization. As previously reported, adipose leptin mRNA was about 5-fold higher in ad-libitum-fed than in fasted mice. By Northern blot analysis (which reflected the entire complement of neurons producing POMC and NPY), POMC mRNA levels were about 5-fold higher in ad-libitum-fed mice than in 48-hour-fasted mice. This result was corroborated in a separate set of mice using an RNAse protection assay, according to which POMC mRNA levels were about 3-fold higher in fed compared to fasted mice. POMC mRNA levels were positively correlated with leptin mRNA levels. Conversely, by Northern blot assay, NPY mRNA levels were about 4-fold higher in 48-hour-fasted mice than in ad-libitum-fed mice. NPY mRNA levels were significantly and inversely correlated with both POMC and leptin mRNA levels. In situ hybridization throughout the extent of the arcuate nucleus corroborated the results with Northern blots: POMC mRNA levels were about 4-fold higher in ad-libitum-fed mice than in fasted mice throughout the extent of the arcuate nucleus, whereas conversely NPY mRNA levels were about 4-fold higher in fasted mice. The effect of fasting on both POMC mRNA and NPY mRNA were similar throughout the extent of the arcuate nucleus, although at about 300 μ from the rostral end of the nucleus, NPY mRNA appeared to be more highly induced by fasting than at other parts of the arcuate nucleus. In a separate group of mice, POMC, NPY, and MCH mRNA levels were assessed by in situ hybridization in ad-libitum-fed and fasted *db/db* and wild-type mice. Corroborating results of the first study, POMC mRNA levels were about 4-fold higher in fed than in fasted wild-type mice, and NPY mRNA levels were about 4-fold higher in fasted than in fed wild-type mice. Similarly, POMC mRNA levels in fed wild-type mice were about 4-fold higher than in fed *db/db* mice (so POMC mRNA levels in fasted wild-type mice were about the same as in fed *db/db* mice). Interestingly, fasting further reduced POMC mRNA levels in *db/db* mice, suggesting that at least some effects of fasting are independent of the leptin pathway. Similarly, NPY mRNA levels were about 4-fold higher in fasted wild-type than in ad-libitum-fed mice, and NPY mRNA levels were also about 4-fold higher in ad-libitum-fed *db/db* mice than in ad-libitum-fed wild-type mice (thus NPY mRNA levels were about the same in ad-libitum-fed *db/db* as in fasted wild-type mice). Furthermore, consistent with the results of POMC mRNA, fasting further elevated the already high levels of NPY mRNA levels in *db/db* mice, again indicating that at least part of the effect of fasting is not due to leptin. MCH mRNA levels were also elevated by fasting and in ad-libitum-fed *db/db* mice compared to wild-type mice. POMC mRNA was negatively correlated with NPY mRNA and with MCH mRNA, which in turn were positively correlated with each other. Finally, as assessed by Northern blot analysis, i.p. injection of leptin (3 μg/BW) into male *ob/ob* mice over 5 days increased POMC mRNA about 2-fold, and decreased NPY mRNA about 30%, but in neither case did leptin replacement at these doses over this period of time completely correct the profile of gene expression. As assessed by in situ hybridization, i.p. injection of leptin (1 μg/g BW) into female *ob/ob* mice over 2 days increased POMC mRNA slightly more than 2-fold in both ad-libitum-fed mice and mice which were fasted for the final 48 h before sacrifice.

Qualitatively, the three papers described above drew the same basic conclusion: that hypothalamic POMC mRNA is decreased in *ob/ob* mice, compared to wild-type, and that leptin can stimulate hypothalamic POMC mRNA in *ob/ob* mice. This general agreement was particularly notable since the first two papers used leptin from a different source than the paper by Mizuno et al. [38]. However, the papers by Schwartz et al. [37] and Mizuno et al. [38] went on to address other points in more detail. For example, Schwartz et al. [37] addressed the role of the leptin receptor by demonstrating that leptin failed to stimulate hypothalamic POMC mRNA in *db/db* mice, in which mutations in the leptin receptor lead to leptin resistance. Furthermore, Schwartz et al. [37] reported that i.c.v. infusion of leptin into normal fasting rats would also stimulate hypothalamic POMC mRNA. However, Schwartz et al. [32] used only in situ hybridization and confined their analysis primarily to the rostral hypothalamus. Thus, the total quantitative regulation of hypothalamic POMC mRNA was not clear. Mizuno et al. [38] used three different methods to quantify POMC mRNA, and also examined several different hypothalamic mRNAs as well as leptin mRNA. Since Northern blots and RNAse protection assay, which included essentially all POMC mRNA from the hypothalamus, generally produced similar results as in situ hybridization, this implied that the effect of leptin was probably not confined to the rostral hypothalamus. However, Mizuno et al. [38] did indicate that the main effect of leptin and fasting were on the lateral aspect of the POMC field, rather than the medial aspect, although this hypothesis was not based on rigorous quantification. By measuring the expression of several different genes and demonstrating significant correlations between them all and with leptin, the study by Mizuno et al. [38] also provided support for the hypothesis that POMC mRNA was roughly equally sensitive to leptin as NPY mRNA and MCH mRNA, a conclusion also supported by direct measurement of the effects of leptin replacement on NPY mRNA. Nevertheless, because Mizuno et al. [38] demonstrated that fasting inhibited hypothalamic POMC mRNA and stimulated NPY mRNA even in *db/db* mice, this suggested leptin is not the sole physiological mediator of nutritional effects on hypothalamic gene expression. Finally, Mizuno et al. [38] also demonstrated that leptin stimulates hypothalamic POMC mRNA in both males and females.

Physiological Significance of the Stimulation of Hypothalamic POMC mRNA by Leptin

Considerable evidence has implicated NPY mRNA in the regulation of body weight [39, 40], and specifically in mediating the effect of leptin [41, 42]. The evidence in support of a role of NPY in regulating body weight includes: (1) NPY stimulates feeding behavior; (2) hypothalamic NPY mRNA is stimulated by fasting; (3) chronic experimental elevation of NPY promotes obesity; (4) *ob/ob* and *db/db* genetically obese mice exhibit elevated NPY mRNA; (5) leptin inhibits hypothalamic NPY mRNA.

The studies described above clearly indicate that leptin stimulates hypothalamic POMC mRNA to roughly the same extent that leptin inhibits NPY mRNA. As described above, similar evidence supports an important role for α-MSH in regulating body weight, although whereas NPY mRNA is anabolic (promoting feeding and weight gain), α-MSH is catabolic (inhibiting feeding and weight gain). Thus: (1) α-MSH and α-MSH agonists inhibit feeding behavior, whereas antagonists to α-MSH stimulate feeding behavior; (2) hypothalamic POMC mRNA (which codes for α-MSH) is inhibited by fasting;

(3) chronic blockade of α-MSH tone (either by overexpression of the *agouti* gene product, overexpression of the *agouti*-related protein gene product, or knocking out the MC4-R α-MSH receptor subtype) promotes obesity; (4) *ob/ob* and *db/db* genetically obese mice exhibit decreased hypothalamic POMC mRNA; (5) leptin stimulates hypothalamic POMC mRNA.

Despite considerable evidence that NPY plays a central role in regulating body weight and promoting obesity, recent studies have required reevaluation of this hypothesis. For example, mice with a targeted deletion of NPY gene show greater susceptibility to seizures, but such mice exhibit no obvious impairments in metabolic or endocrine regulation [43–45]. In particular, NPY knockout mice appear to respond normally to leptin [44]. Nevertheless, the obesity of *ob/ob* mice is somewhat attenuated in the absence of the NPY gene [45], suggesting that NPY does mediate some effects of leptin (or at least leptin deprivation). Using a different approach, Bergen et al. [46] demonstrated that production of obesity by i.p. injection of gold-thioglucose (GTG) was not demonstrably attenuated by prior elimination of hypothalamic NPY by neonatal treatment with monosodium glutamate. Therefore, although a large body of evidence implicates NPY in the regulation of body weight and in obesity, the precise nature of NPY's role remains to be clarified.

Similar studies have also demonstrated that the role of hypothalamic POMC mRNA, like NPY, remains to be clarified. Although no α-MSH knockout mice have been reported, the *agouti* mouse may be considered in some ways similar to an α-MSH knockout, since melanocortin signaling is highly compromised. Using *agouti* mice, Boston et al. [47] carried out a study analogous to that of Erickson et al. [45], and assessed the effect of the *agouti* mutation (which blocks α-MSH receptors) on the obesity produced in leptin-deficient *ob/ob* mice. Although *agouti* mice were relatively insensitive to the effects of leptin, offspring of *ob/ob* and *agouti* mice were highly sensitive to leptin. Furthermore, these offspring exhibited an even more exaggerated obese phenotype than *ob/ob* or *agouti* mice alone. The authors concluded that the obesity of *agouti* mice is leptin-independent, and that *agouti* mice are only insensitive to leptin because of down-regulation secondary to chronic leptin elevation. One interpretation of this conclusion is that, since the blockade of the melanocortin system did not attenuate the effect of leptin (in leptin-deficient mice), the melanocortin system did not mediate any major effects of leptin (similar to the result in the NPY knockout mice). Conversely, this interpretation would suggest that obesity resulting from interfering with the melanocortin system is independent of leptin.

Seeley et al. [48] assessed the role of the melanocortin system in mediating effects of leptin using a pharmacological approach. As described above, Fan et al. [13] had characterized a nonpeptide compound, SHU9119, which antagonizes both the MC3-R and MC4-R, though like the *agouti* protein the antagonism of the MC4-R was greater. When Seeley et al. [48] infused this compound into the third ventricle, the effects of an i.c.v. infusion of leptin on food intake and induction of c-fos were significantly attenuated. Thus at least in this short-term paradigm the melanocortinergic pathway appears to mediate at least some effects of leptin.

Leptin and β-Endorphin

Most forms of obesity, particularly obesity due to leptin deficiency and leptin resistance, are associated with elevated glucocorticoids. Interestingly, however, the obesity of the *agouti* mouse is not associated with elevated glucocorticoids. This latter observation suggests that α-MSH does not mediate the effects of leptin to reduce glucocorticoid secretion. However, this does not rule out the possibility that other POMC products besides α-MSH may mediate some effects of leptin. In particular, β-endorphin, acting on hypothalamic neurons (presumably corticotropin-releasing-hormone neurons) has long been known to decrease secretion of glucocorticoids. Therefore it is plausible that the effect of leptin to reduce CRF release may be mediated through elevation of β-endorphin. This would potentially explain why *agouti* is one of the only models of obesity not associated with elevated glucocorticoids: it is also one of the few models of obesity in which POMC synthesis and processing is intact, thus being one of the few models in which hypothalamic β-endorphin is probably not decreased. The general relationship between hypothalamic β-endorphin and elevated glucocorticoids in obesity remains to be determined.

POMC and Obesity

Virtually every form of obesity for which the biological basis is known exhibits decreased production, processing, or sensitivity to POMC products. Decreased POMC mRNA in the hypothalamus is associated with obesity produced by the canine distemper virus [49] and obesity produced by deletion of the gene for the basic helix loop-helix Nhlh2 transcription factor [50]. Furthermore, the process of aging, particularly in rats, is associated with the development of obesity, and hypothalamic POMC mRNA decreases during aging [51]. As described above, obesity due to leptin insufficiency and leptin resistance is also associated with decreased hypothalamic POMC mRNA [36–38]. Furthermore, the obesity of the fat/fat mouse is due to a mutation in the carboxypeptidase E gene [52], which leads to impaired ability to sort and/or process POMC, among other peptides [53].

Of particular interest, recent studies have revealed the genes responsible for four forms of hereditary obesity in humans, and all are either known or can be assumed to be associated with impairments in the POMC system. Two of these involve mutations in either leptin [54], or the leptin receptor [55], and, based on studies in rodents [36–38], it seems likely that hypothalamic POMC would be decreased in such individuals. Furthermore, heterozygote compound mutations in the prohormone convertase 1 (PC1) gene produce human obesity in association with impaired processing of POMC and insulin [56]. Thus, mutation in the PC1 gene produces a phenotype in humans similar to the phenotype in mice produced by a mutation in the carboxypeptidase gene [52]. Finally, and most convincingly, it has recently been demonstrated that different mutations in the POMC gene itself, which lead to a loss of POMC-derived peptides, cause obesity in humans [57].

Thus, impairments in the POMC system can directly cause obesity, and such impairments are associated with most, if not all forms of obesity. Most forms of obesity are also associated with elevated leptin levels, which is assumed to imply leptin resistance. However, in view of the evidence that the main effects of leptin are not mediated through POMC [47], the hypothesis must be entertained that impairments in the POMC system are more generally a cause of obesity, and that elevated leptin, and/or leptin resistance, arises as a secondary consequence to the impairments in the POMC system. If so, this may suggest that the POMC system is at least as promising a system for intervention as the leptin system.

Conclusions

Most, if not all, forms of obesity are associated with impairments in the synthesis, processing, or sensitivity to products derived from hypothalamic POMC. In particular, obesity due to leptin insufficiency or leptin resistance is associated with decreased hypothalamic POMC mRNA, and, conversely, leptin stimulates hypothalamic POMC mRNA. The most likely product of POMC which acts to reduce feeding and prevent obesity is α-MSH. Nevertheless, it appears likely that many, if not most effects of leptin are independent of α-MSH, since *agouti* mice with leptin insufficiency are highly sensitive to the weight-reducing effects of leptin, although other effects of leptin, such as leptin-mediated reduction of glucocorticoids, may involve the stimulation of hypothalamic β-endorphin. Conversely, some effects of fasting on POMC mRNA are independent of leptin, since effects of fasting on POMC mRNA can be observed even in *db/db* mice. Therefore, it seems clear that though the leptin system and the POMC system overlap, they are also to some extent independent. In view of the many forms of obesity known to be caused by impairments in the POMC system, this system remains an important potential target for intervention, independent of the leptin system.

References

1 Bultman SJ, Michaud EJ, Woychik RP: Molecular characterization of the mouse agouti locus. Cell 1992;71:1195–1204.

2 Bultman SJ, Russell LB, Gutierrez-Espeleta GA, Woychik RP: Molecular characterization of a region of DNA associated with mutations at the agouti locus in the mouse. Proc Natl Acad Sci USA 1991;88:8062–8066.

3 Miller MW, Duhl DMJ, Vrieling H, Cordes SP, Ollmann MM, Winkes BM, Barsh GS: Cloning of the mouse agouti gene predicts a secrete protein ubiquitously expressed in mice carrying the lethal yellow mutation. Genes Dev 1993;7:454–467.

4 Hearing VJ, Tsukamoto K: Enyzmatic control of pigmentation in mammals. FASEB J 1991;5: 2902–2909.

5 Geschwind II: Change in hair color in mice induced by the injection of α-MSH. Endocrinology 1966;79:1165–1167.

6 Tamate HB, Takeuchi T: Action of the e locus of mice in the response of phaemelanic hair follicles to α melanocyte stimulating hormone in vitro. Science 1984;224:1241–1242.

7 Takeuchi T, Kobunai T, Yamamoto H: Genetic control of signal transduction in mouse melanocytes. J Invest Dermatol 1989;92:2395–2425.

8 Mountjoy KG, Robbins LS, Mortrud MT, Cone RD: The cloning of a family of genes that encode the melanocortin receptors. Science 1992;257:1248–1251.

9 Robbins LS, Nadeau JH, Johnson KR, Kelly MA, Roselli-Rehfuss L, Baack E, Mountjoy KG, Cone RG: Pigmentation phenotypes of variant extension locus alleles result from point mutations that alter MSH receptor function. Cell 1993;72:827–834.

10 Lu D, Willard D, Patel IR, Kadwell S, Overton L, Kost T, Luther M, Chen W, Woychik RP, Wilkison WO, Cone RD: Agouti protein is an antagonist of the melanocyte-stimulating-hormone receptor. Nature 1994;371:799–802.

11 Mountjoy KG, Mortrud MT, Low MJ, Simerly RB, Cone RD: Localization of the melanocortin-4 receptor (MC4-R) in neuroendocrine and autonomic control circuits in the brain. Mol Endocrinol 1994;8:1298–1308.

12 Huszar D, Lynch CA, Fairchild-Huntress V, Dunmore JH, Fang Q, Berkemeier LR, Gu W, Kesterson RA, Boston BA, Cone RD, Smith FJ, Campfield LA, Burn P, Lee F: Targeted disruption of the melanocortin-4 receptor results in obesity in mice. Cell 1997;88:131–141.

13 Fan W, Boston BA, Kesterson RA, Hruby VJ, Cone RD: Role of melanocortinergic neurons in feeding and the agouti obesity syndrome. Nature 1997;385:165–168.

14 Grandison L, Guidotti A: Stimulation of food intake by muscimol and beta endorphin. Neuropharmacology 1977;16:533–536.

15 Levine AS, Morley JE, Gosnell BA, Billington CJ, Bartness TJ: Opioids and consummatory behaviors. Brain Res Bull 1985;14:663–672.

16 Roberts JL, Seeburg PH, Shine J, Herbert E, Baxter JD, Goodman HM: Corticotropin and beta-endorphin: Construction and analysis of recombinant DNA complementary to mRNA for the common precursor. Proc Natl Acad Sci USA 1979;76:2153–2157.

17 Gee CE, Chen CL, Roberts JL, Thompson R, Watson SJ: Identification of proopiomelanocortin neurones in rat hypothalamus by in situ cDNA-mRNA hybridization. Nature 1983;306:374–376.

18 Brady LS, Smith MA, Gold PW, Herkenham M: Altered expression of hypothalamic neuropeptide mRNAs in food-restricted and food-deprived rats. Neuroendocrinology 1990;52:441–447.

19 Bergendahl M, Wiemann JN, Clifton DK, Huhtaniemi I, Steiner RA: Short-term starvation decreases POMC mRNA but does not alter GnRH mRNA in the brain of adult male rats. Neuroendocrinology 1992;56:913–920.
20 Kim EM, Welch CC, Grace MK, Billington CJ, Levine AS: Chronic food restriction and acute food deprivation decrease mRNA levels of opioid peptides in arcuate nucleus. Am J Physiol 1996; 270:R1019–R1024.
21 Jhanwar-Uniyal M, Chua SC Jr: Critical effects of aging and nutritional state on hypothalamic neuropeptide Y and galanin gene expression in lean and genetically obese Zucker rats. Brain Res Mol Brain Res 1993;19:195–202.
22 Schwartz MW, Sipols AJ, Grubin CE, Baskin DG: Differential effect of fasting on hypothalamic expression of genes encoding neuropeptide Y, galanin, and glutamic acid decarboxylase. Brain Res Bull 1993;31:361–367.
23 Vergoni AV, Poggioli R, Bertonlini A: Corticotropin inhibits food intake in rats. Neuropeptides 1986;7:153–158.
24 Poggioli R, Vergoni AV, Bertolini A: ACTH-(1-24) and α-MSH antagonize feeding behavior stimulated by kappa opiate agonists. Peptides 1986;7:843–848.
25 Tsujii S, Bray GA: Acetylation alters the feeding response to MSH and β-endorphin. Brain Res Bull 1989;23:165–169.
26 Fong TM, Mao C, MacNeil T, Kalyani R, Smith T, Weinberg D, Tota MR, Van der Ploeg LH: ART (protein product of agouti-related transcript) as an antagonist of MC-3 and MC-4 receptors. Biochem Biophys Res Commun 1997;237:629–631.
27 Zhang Y, Proenca R, Maffe M, Barone M, Leopold L, Friedman JM: Positional cloning of the mouse obese gene and its human homologue. Nature 1994;372:425–432.
28 Halaas JL, Gajiwala KS, Maffei M, Cohen SL, Chait BT, Rabinowitz D, Lallone RL, Burley SK, Friedman JM: Weight-reducing effects of the plasma protein encoded by the obese gene. Science 1995;269:543–546.
29 Mizuno TM, Bergen H, Funabashi T, Kleopoulos SP, Zhong YG, Bauman WA, Mobbs CV: Obese gene expression: Reduction by fasting and stimulation by insulin and glucose in lean mice, and persistent elevation in acquired (diet-induced) and genetic (yellow agouti) obesity. Proc Natl Acad Sci USA 1996;93:3434–3438.
30 Saladin R, De Vos P, Guerre-Millo M, Leturque A, Girard J, Staels B, Auwerx J: Transient increase in obese gene expression after food intake or insulin administration. Nature 1995;377:527–529.
31 Ahima RS, Prabakaran D, Mantzoros C, Qu D, Lowell B, Maratos-Flier E, Flier JS: Role of leptin in the neuroendocrine response to fasting. Nature 1996;382:250–252.
32 Schwartz MW, Seeley RJ, Campfield CA, Burn P, Baskin DG: Identification of targets of leptin action in rat hypothalamus. J Clin Invest 1996;98:1101–1106.
33 Tartaglia LA, Dembski M, Weng X, Deng N, Culpepper J, Devos R, Richards GJ, Campfield LA, Clark FT, Deeds J, et al: Identification and expression cloning of a leptin receptor, OB-R. Cell 1995;83:1263–1271.
34 Lee GH, Proenca R, Montez JM, Carroll KM, Darvishzadeh JG, Lee JI, Friedman JM: Abnormal splicing of the leptin receptor in diabetic mice. Nature 1996;379:632–635.
35 Cheung CC, Clifton DK, Steiner RA: Proopiomelanocortin neurons are direct targets for leptin in the hypothalamus. Endocrinology 1997;138:4489–4492.
36 Thornton JE, Cheung CC, Clifton DK, Steiner RA: Regulation of hypothalamic proopiomelanocortin mRNA by leptin in ob/ob mice. Endocrinology 1997;138:5063–5066.
37 Schwartz MW, Seeley RJ, Woods SC, Weigle DS, Campfield LA, Burn P, Baskin DG: Leptin increases hypothalamic pro-opiomelanocortin mRNA expression in the rostral arcuate nucleus. Diabetes 1997;46:2119–2123.
38 Mizuno TM, Kleopoulos SP, Bergen HT, Roberts JL, Priest CA, Mobbs CV: Hypothalamic pro-opiomelanocortin mRNA is reduced by fasting in ob/ob and db/db mice, but is stimulated by leptin. Diabetes 1998;47:294–297.
39 Billington CJ, Briggs JE, Harker S, Grace M, Levine AS: Neuropeptide Y in hypothalamic paraventricular nucleus: A center coordinating energy metabolism. Am J Physiol 1994;266:R1765–R1770.

40 Levine As, Billington CJ: Why do we eat? A neural systems approach. Annu Rev Nutr 1997;17: 597–619.
41 Stephens TW, Basinski M, Bristow PK, Bue-Valleskey JM, Burgett SG, Craft L, Hale J, Hoffmann J, Hsiung HM, Kriauciunas A, Mackellar W, Rosteck PR Jr, Schoner B, Smith D, Tinsley FC, Zhang X-Y, Heiman M: The role of neuropeptide Y in the antiobesity action of the obese gene product. Nature 1995;377:530–532.
42 Schwartz MW, Baskin DG, Bukowski TR, Kuijper JL, Foster D, Lasser G, Prunkard DE, Porte D Jr, Woods SC, Seeley RJ, Weigle DS: Specificity of leptin action on elevated blood glucose levels and hypothalamic neuropeptide Y gene expression in ob/ob mice. Diabetes 1996;45:531–535.
43 Erickson JC, Ahima RS, Hollopeter G, Flier JS, Palmiter RD: Endocrine function of neuropeptide Y knockout mice. Regul Pept 1997;70:199–202.
44 Erickson JC, Clegg KE, Palmiter RD: Sensitivity to leptin and susceptibility to seizures of mice lacking neuropeptide Y. Nature 1996;381:415–421.
45 Erickson JC, Hollopeter G, Palmiter RD: Attenuation of the obesity syndrome of ob/ob mice by the loss of neuropeptide Y. Science 1996;274:1704–1707.
46 Bergen H, Mizuno TM, Taylor J, Mobbs CV: Hyperphagia and weight gain after gold-thioglucose: Relation to hypothalamic neuropeptide Y and proopiomelanocorticotropin. Endocrinology 1998; 139:4483–4488.
47 Boston BA, Blaydon KM, Varnerin J, Cone RD: Independent and additive effects of central POMC and leptin pathways on murine obesity. Science 1997;278:1641–1644.
48 Seeley RJ, Yagaloff KA, Fisher SL, Burn P, Thiele TE, van Dijk G, Baskin DG, Schwartz MW: Melanocortin receptors in leptin effects (letter). Nature 1997;390:349.
49 Nagashima K, Zabriskie JB, Lyons MJ: Virus-induced obesity in mice: Association with a hypothalamic lesion. J Neuropathol Exp Neurol 1992;51:101–109.
50 Good DJ, Porter FD, Mahon KA, Parlow AF, Westphal H, Kirsch IR: Hypogonadism and obesity in mice with a targeted deletion of the Nhlh2 gene. Nat Genet 1997;15:397–401.
51 Lloyd JM, Scarbrough K, Weiland NG, Wise PM: Age-related changes in proopiomelanocortin (POMC) gene expression in the periarcuate region of ovariectomized rats. Endocrinology 1991;129: 1896–1902.
52 Naggert JK, Fricker LD, Varlamov O, Nishina PM, Rouille Y, Steiner DF, Carroll RJ, Paigen BJ, Leiter EH: Hyperproinsulinaemia in obese fat/fat mice associated with a carboxypeptidase E mutation which reduces enzyme activity. Nat Genet 1995;10:135–142.
53 Cool DR, Normant E, Shen F, Chen HC, Pannell L, Zhang Y, Loh YP: Carboxypeptidase E is a regulated secretory pathway sorting receptor: Genetic obliteration leads to endocrine disorders in Cpe(fat) mice. Cell 1997;88:73–83.
54 Montague CT, Farooqi IS, Whitehead JP, Soos MA, Rau H, Wareham NJ, Sewter CP, Digby JE, Mohammed SN, Hurst JA, Cheetham CH, Earley AR, Barnett AH, Prins JB, O'Rahilly S: Congenital leptin deficiency is associated with severe early-onset obesity in humans. Nature 1997;387:903–908.
55 Clement K, Vaisse C, Lahlou N, Cabrol S, Pelloux V, Cassuto D, Gourmelen M, Dina C, Chambaz J, Lacorte JM, Basdevant A, Bougneres P, Lebouc Y, Froguel P, Guy-Grand B: A mutation in the human leptin receptor gene causes obesity and pituitary dysfunction. Nature 1998;392:398–401.
56 Jackson RS, Creemers JWM, Ohagi S, Raffin-Sanson ML, Sanders L, Montague CT, Hutton JO, O'Rahilly S: Obesity and impaired prohormone processing associated with mutations in the human prohormone convertase 1 gene. Nat Genet 1997;16:303–306.
57 Krude H, Biebermann H, Luck W, Horn R, Brabant G, Gruters A: Severe early-onset obesity, adrenal insufficiency and red hair pigmentation caused by POMC mutations in humans. Nat Genet 1997;19:155–157.

Charles V. Mobbs, Dr. Arthur Fishberg Research Center for Neurobiology,
Mount Sinai School of Medicine, Box 1065,
One Gustav L. Levy Place, New York, NY 100291-6574 (USA)
Tel. +1 212 824 8738, Fax +1 212 849 2510, E-Mail mobbsc@alum.mit.edu

Ur E (ed): Neuroendocrinology of Leptin.
Front Horm Res. Basel, Karger, 2000, vol 26, pp 71–86

Hypothalamic Neuropeptide Y and Its Neuroendocrine Regulation by Leptin

Peter S. Widdowson, John P.H. Wilding

Diabetes and Endocrinology Research Unit, Department of Medecine,
Liverpool University, Liverpool, UK

The discovery of the lipostat hormone, leptin, has significantly increased the understanding of how peripheral signals interact with centres in the brain to regulate body fat content via changes in appetite and thermogenesis. Although leptin, which is derived from adipose tissue and is released into the circulatory system in direct proportion to body fat mass, interacts with numerous neurochemical signalling systems, including the α-MSH/melanocortin/MC4 system, glucagon-like peptide-1 (GLP-1) and corticortrophin-releasing factor (CRF), the most widely studied to date is the interaction of leptin with neurones that utilise neuropeptide Y (NPY) as a neurotransmitter. These effects may well be important in mediating the effects of leptin on food intake, but evidence is also emerging that thc neuroendocrine effects of leptin on the NPY system are also important in the regulation of reproductive function in both puberty and starvation.

Neuropeptide Y

Neuropeptide Y (NPY) is a 36-amino acid peptide (named because of the presence of tyrosine at both C- and N-terminals of the peptide) which was first isolated in 1983. It is found throughout the central and peripheral nervous systems, where it acts as a neurotransmitter. In the central nervous system NPY is found in the cortex, hippocampus and many other brain regions, but it is present in particularly high abundance in the hypothalamus of mammals, including humans. Hypothalamic NPY is derived from two sources, intrahypothalamic, from NPYergic neurones of the arcuate nucleus (ARC) and from extrahypothalamic nuclei located primarily in the medulla oblongata which transport and release NPY together with other monoamines, specifically noradrenaline and serotonin. Most hypothalamic NPY is derived from the arcuate nucleus, and from this site neuronal projections pass dorsally and anteriorly to innervate the hypothalamic paraventricular (PVN) and dorsomedial nuclei (DMH). There are also minor projections to other hypothalamic nuclei from NPY containing ARC-derived cells [1].

For over 15 years, it has been known that NPY has important effects on energy balance. Injection of NPY into the brain, particularly the perifornical and paraventricular hypothalamus, results in a rapid increase in food consumption with a latency of onset of around 20 min and a duration of action of about 4 h. This stimulation of food intake is so powerful that it appears to override all the usual satiety signals, such that an animal may consume up to 50% of its usual daily intake in 2 h, even when the peptide is given to satiated animals. NPY is unique amongst neurotransmitters known to stimulate food consumption, such as noradrenaline, galanin and the recently discovered orexin, in that the hyperphagia is long-lasting and sustained following multiple injections, so that chronic administration rapidly leads to obesity, with associated hyperinsulinaemia and sustained elevations in plasma insulin concentrations, resulting in significant increases in fat mass and body weight over a relatively short period of 10 days [2]. Furthermore, the increase in food consumption is accompanied by a reduction in the sympathetic outflow to heat-generating brown adipose tissue (BAT), resulting in reductions in BAT thermogenesis which may further contribute to weight gain [3].

NPY as a Physiological Mediator of Food Intake

The ARC NPY neurons are highly sensitive to changes in nutritional status. NPY mRNA expression in the ARC rises within 24 h of food deprivation, and this is accompanied by increased NPY concentrations in the ARC and sites of the NPY neuronal projections, such as the PVN. Levels return rapidly to normal on refeeding. This increase in concentration is accompanied by increased NPY release from the region of the PVN, as has been demonstrated by push-pull cannulation and microdialysis sampling [4, 5]. A physiological role for the endogenous high NPY levels in controlling food intake after food deprivation has been demonstrated using neutralising monoclonal antibodies, and more recently by using antisense oligonucleotides to inhibit endogenous NPY synthesis [6, 7]. It has, however, proven more difficult to establish a role for endogenous NPY in regulating day-to-day food intake, for example, the NPY knockout mouse exhibits normal food intake under most circumstances, although preliminary reports have suggested that blockade of synthesis of the NPY Y5 receptor (see below) may attenuate normal feeding at the onset of the dark phase in rats.

Other Central Effects of Neuropeptide Y

Centrally administered NPY has been shown to have a number of other neuroendocrine effects which may be relevant to leptin action. These include stimulation of insulin and corticosterone secretion, and modulation of luteinising hormone (LH) secretion. Acute injection of neuropeptide Y into the third ventricle, hypothalamus, or fourth ventricle results in a prompt increase in circulating insulin, even in the absence of food [8]. Chronic NPY adminstration results in hyperinsulinaemia and insulin resistance, even when animals are pair-fed to compensate for any effects due to weight gain [2]. Corticosterone secretion is also acutely increased by NPY. NPY injected into the hypothalamus inhibits GnRH secretion, and thus LH secretion, in most species studied, and it has been suggested that elevated NPY in starvation, and during lactation, may partly be responsible for the inhibition of LH secretion which occurs under these conditions [9, 10].

NPY Receptors

Considerable interest has been shown in characterising and identifying the receptor subtype that mediates the hyperphagic and anti-thermogenic actions of NPY, as selective drugs with high affinity which block these receptors have therapeutic potential in the treatment of obesity. Pharmacological and molecular cloning techniques have demonstrated that the actions of NPY, both in the periphery and brain could be explained by interactions at multiple NPY subtypes, each with distinct pharmacology (fig. 1). Early characterisation of these NPY receptor subtypes was difficult owing in part to the absence of selective antagonists, with most of the evidence for multiple receptors being derived from studies of C-terminal NPY fragments and modified NPY amino acid sequences. These early pharmacological experiments identified two classes of NPY receptors called Y1 and Y2 subtypes. The Y1 subtype required the entire NPY sequence for full activity since the C-terminal fragments NPY(2-26), NPY(3-36) and NPY(13-36) showed much reduced activity or were inactive, but Pro^{34} substituted NPY peptides were equipotent to NPY at this receptor. In contrast, Y2 receptors exhibited similar potency of C-terminal NPY fragments as compared to the whole NPY sequence, whilst Pro^{34}-substituted peptides were inactive. The structurally related hormone, peptide YY (PYY), which is produced in the intestinal tract is equipotent, or slightly more potent than NPY at both Y1 and Y2 receptors. Both Y1 and Y2 receptors have been cloned from human and rat tissues and cell lines and have been demonstrated to exhibit a pharmacological profile equivalent to that predicted with NPY peptide fragments. These Y1 and Y2 receptors are widely distributed in rodent brains, as revealed using radioligands such as [Leu^{31}, Pro^{34}]PYY for Y1 receptors or PYY(3-36) for Y2 receptors. Y1 receptors are particularly well expressed in the outer layers of the cerebral

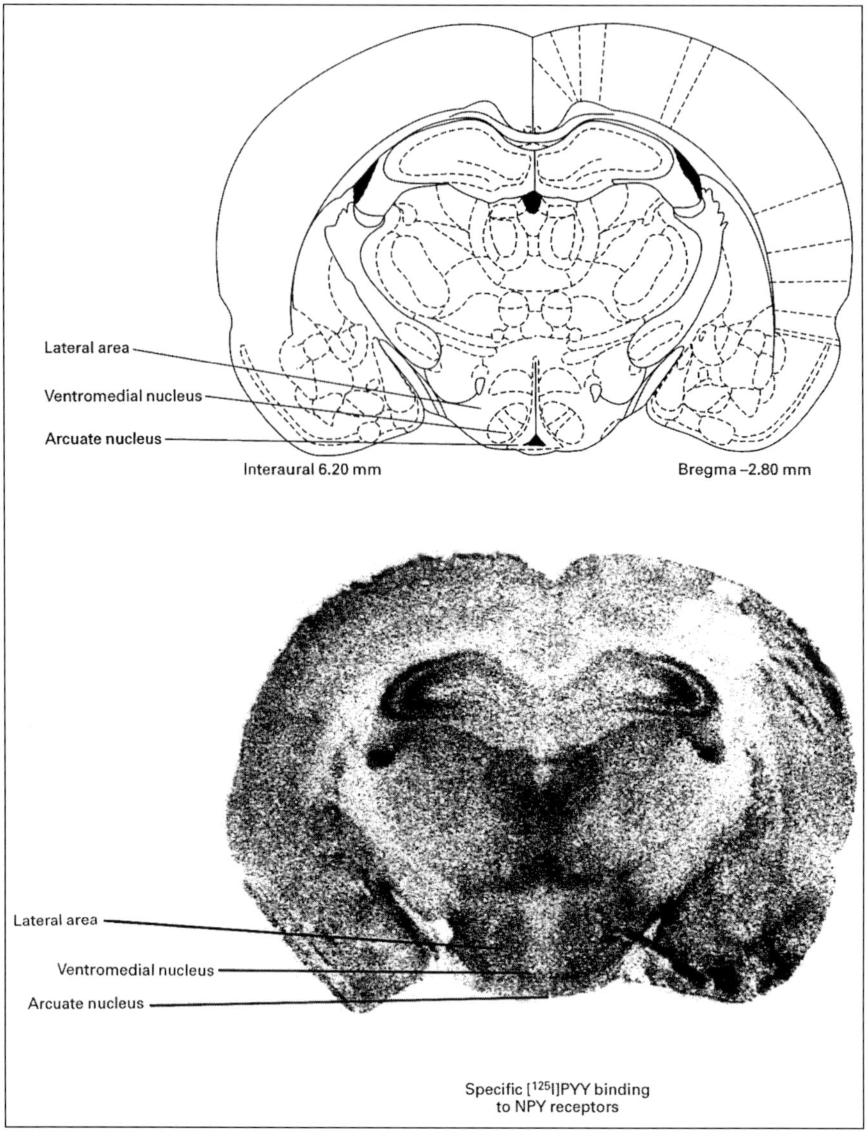

Fig. 1. NPY binding in the hypothalamus.

cortex and midline thalamic nuclei whilst Y2 receptors predominate in the hippocampus and hypothalamus [14]. Pharmacological studies have also predicted other NPY receptors including the Y3 receptor which mediates NPY's hypertensive effects when injected into the medulla oblongata, but is less sensitive to PYY. To date the Y3 receptor has not yet been cloned. Receptors for rat and human pancreatic polypeptides (PPs) have also been isolated and cloned and named Y4 receptors since they also exhibit nanomolar affinity for NPY and PYY. Y4 receptors are also expressed in the brain, but show a more limited distribution to Y1 and Y2 receptors, as revealed using rat PP.

A novel receptor may mediate the hyperphagic effects of NPY since the Y1 agonist NPY is equipotent to NPY, as is the C-terminal NPY fragment, NPY(2-36) which displays a much lower affinity at Y1 receptors, whilst the Y2 ligand NPY(13-36) is almost inactive. Hence, the NPY 'feeding' receptor was demonstrated to exhibit a Y1-like profile, but was significantly different to the previously characterised Y1 receptor [15]. Further experiments confirmed and extended the data supporting a role for a novel NPY receptor in the feeding response. Subsequently, a novel NPY receptor was identified from rat hypothalamus exhibiting the preferred pharmacological profile of the 'feeding' receptor and was called Y5 [16]. Y5 receptor synthesis has been localised to a number of brain regions using in situ hybridisation and regions with high synthesis include the hypothalamic PVN, thalamic centromedial and reuniens nuclei and the dentate gyrus of the hippocampus (fig. 1). The human homologue of the rat Y5 receptor has now also been cloned and demonstrated to be produced in the brain. Lastly, a sixth NPY receptor was identified in mouse brain and called Y6, with a pharmacology almost identical to that of the rat and human Y5 receptor except that NPY has much lower affinity for the Y6 receptors, as compared to Y5 receptors. The Y6 receptor does not appear to play a role in the NPY-induced hyperphagia in the rat since it is not expressed in either rat or human tissues.

Many high-affinity non-peptide selective NPY receptor antagonists have begun to be described, displaying selectivity for Y1 and Y5 receptors which has led to some debate as to which receptor may mediate the NPY feeding response, despite results obtained with N-terminal NPY fragments. Support for a role for the Y5 receptor has come from experiments using anti-sense knockout of the Y5 receptor and demonstrating a subsequent reduction in food intake following both NPY injections and overnight starvation [17]. Blockage of the Y5 receptor with the high-affinity selective antagonist CGP71683A (Novartis) in *ob/ob* mice and obese (*fa/fa*) Zucker rats again strongly suggests a role for Y5 receptors in mediating the hyperphagic response to endogenous or exogenous NPY. However, high-affinity non-peptide Y1 antagonists, which do not have any significant affinity for Y5 receptors, such as BIBO3304 (Boehringer Ingleheim) and LY366,337 (Lilly), are also effective in reducing food consumption in *ob/ob* mice and *fa/fa* rats. Furthermore, a role for Y5 receptors in the NPY-induced feeding response has been questioned by reports that Y5 antagonists, such as L-154, XX (Merck/Banyu Pharmaceuticals) are unable to inhibit NPY-induced feeding nor significantly attenuate

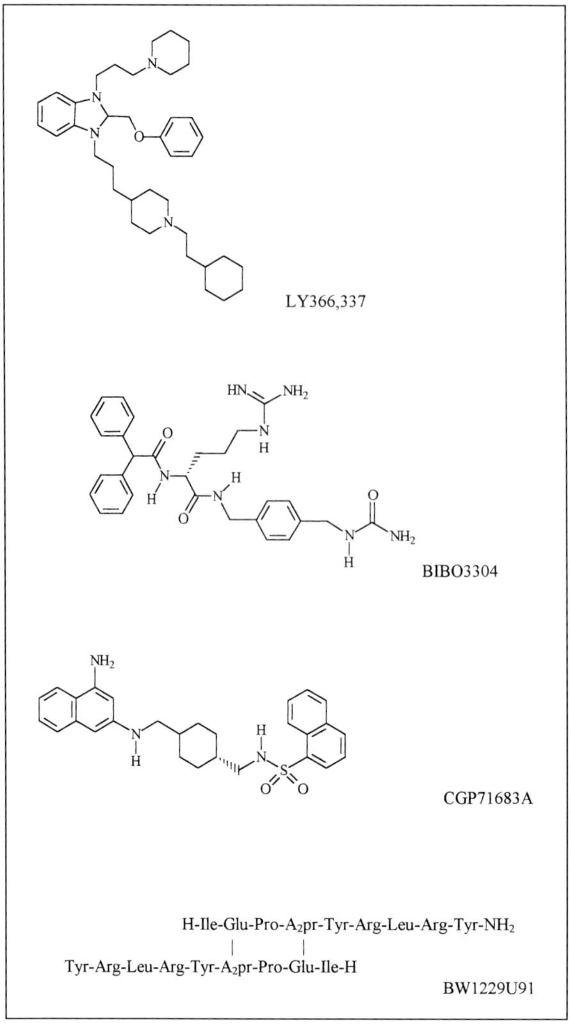

Fig. 2. Structure of selected NPY antagonists.

the hyperphagia in obese Zucker rats (fig. 2). The final outcome of this debate is uncertain, but at present it seems likely that both Y1 and Y5 receptors are involved, though a role for a third, as yet undiscovered receptor is still possible.

NPY in Animal Models of Obesity

The powerful orexigenic effects of NPY suggested to many investigators that it could be a possible mediator of obesity in some or all of the monogenic animal models of obesity. It is now clear that NPY expression is increased in many, but not all of these strains, particularly the *ob/ob* mouse (the only model of leptin deficiency), the *db/db* mouse, the Fatty Zucker rat and the JCR-LA corpulent rat (all models with leptin receptor defects). Interestingly, reduced or unchanged NPY expression is found in the arcuate nucleus of the MC4 knockout mouse, a model in which leptin is not involved in the pathogenesis of the obesity syndrome, but increased expression is seen in the dorsomedial hypothalamus in this model. In models of dietary obesity, produced by feeding animals varied, high fat, or highly palatable diets, NPY expression has generally been found to be unchanged or even slightly decreased, and this is supported by the recent observation of NPY receptor downregulation in this model [14].

Leptin

The Discovery of Leptin

For many decades it was proposed that a physiological system of regulating body weight to within 1% of ideal set-limits exists, as epidemiological studies before 1980 showed only modest increases in average body weight in a number of societies. This led to the proposal that there is a strict control of energy intake as food, and energy expenditure including physical exercise and non-shivering thermogenesis. These two sides of the energy equation are highly co-ordinated to maintain a constant level of body fat and weight. The energy balance regulating system requires that afferent signals would pass from energy stores (fat deposits) to the central nervous system to inform the brain of the status of these stores so that adjustments could be made to change both the energy intake and expenditure in a particular direction. The lipostat hypothesis, as it became known, postulated that an unknown signalling mechanism existed to inform the brain of changes in body fat mass, but it was not known until the mid-1990s whether this was through hormonal influences or afferent signalling via the peripheral nervous system.

Parabiosis studies in genetically obese mice, which result from autosomal recessive inherited mutations, strongly suggested the presence of a blood-borne factor which could regulate food intake and energy expenditure. It was suggested from these early experiments that the *ob/ob* mouse was deficient in such a factor, and that the phenotypically identical *db/db* mouse had a defect in its receptor. In 1994, Friedman and colleagues were able to identify the genetic defect responsible for the phenotype of *ob/ob* mice, using a positional cloning approach [18]. The newly discovered ob gene was found to encode a previously unknown 167-amino acid, 16-kD protein. Two co-iosogenic *ob/ob* mouse strains were demonstrated to result in a defective synthesis of the protein product of the *ob* gene via a C to T mutation resulting in a change from arginine[105] to a stop codon, or in the other case, a presumed deletion in the promotor region since these mice failed to synthesise any *ob* mRNA [18]. Northern blotting demonstrated that the *ob* protein was exclusively produced in adipose tissue and secreted into blood of normal mice and was subsequently called 'leptin' derived from the Greek word 'leptos' meaning thin. It was then discovered that leptin release was regulated by a number of hormonal influences, including levels of insulin and corticosterone, but the most overriding factor governing plasma leptin concentrations, in laboratory animals and humans, is body fat mass [9]. Feeding rodents with a high-fat diet, which results in increases in fat depots, is also paralleled by increased plasma leptin concentrations. Conversely, weight reduction, by starvation, diabetes or dietary restriction, leads to reductions in plasma leptin concentrations in response to reductions in body fat mass [19].

Effect of Leptin on the Brain

When administered to mice, recombinant leptin was demonstrated to have a profound effect on energy homeostasis. Daily intraperitoneal injections of leptin to normal or *ob/ob* mice significantly reduced their body weight over time. The *ob/ob* mice were more sensitive to the effects of leptin than normal mice with up to 40% body weight reductions observed over a 30-day experimental period [20]. Body composition analysis demonstrated that the reduction in body weight was accounted for by exclusive reductions in body fat mass, with no change to the lean body mass. The reduction in body fat by chronic leptin treatment was produced by a combination of reduced food intake and increased energy expenditure. Food intake in the normally hyperphagic *ob/ob* mice could be reduced to 60% of saline treated animals by chronic leptin treatment. Leptin-treated *ob/ob* mice shed 30% more fat than pair-fed animals as a result of increased energy expenditure. Further experiments showed that the anorexic and thermogenic effects of systemic leptin could be mimicked by intracerebroventricular administration. Direct administration of leptin into the brain produced more potent effects than those effects observed afier intraperitoneal or subcutaneous leptin injections [21, 22] Chronic leptin injections to *ob/ob* mice also significantly improved their metabolic status since there was significant reductions in plasma insulin and glucose concentrations, which fell to values of normal weight mice. Corticosterone concentrations were also significantly reduced in leptin-treated *ob/ob* mice, but the levels did not reach those of control-lean mice, even after a prolonged period of leptin administration [22].

Leptin Receptor

Because the effects of leptin on body weight were most pronounced when the protein was injected into the brain, it was concluded that leptin must produce its actions directly on the central nervous system. The identification of a high-affinity leptin-binding protein (Ob-R) cloned from mouse choroid plexus, which exhibits a high density of leptin binding [23] provided the final piece of evidence to support the hypothesis for the lipostat signalling from adipose tissue to the brain. Alternate splicing of the mRNA coding for the leptin receptor led to several transcripts being identified (Ob-Ra, Ob-Rb, Ob-Rc, Ob-Rd, Ob-Re), with varying length of the final protein. All the transcripts retain a leptin-binding domain, and all but one transcript (Ob-Re) has a single transmembrane domain suggesting that the Ob-Re transcript may act as a soluble leptin-binding transporter. The amino acid sequence of the leptin receptor resembles the class I cytokine receptor family, which includes the gp130 signal transducing component of the interleukin-6 receptor [23]. Only one of the leptin receptors (Ob-Rb) has an intracellular domain of any significant length (302 amino acids) that may couple to intracellular second messenger systems. The similarity of the Ob-Rb receptor isoform to other class I cytokine receptors predicted that intracellular signal transduction probably involves cytoplasmic protein kinases, particularly those of the JAK (Janus kinase) and STAT (signal transducers and activators of transcription) family. In situ hybridisation studies have demonstrated a selective expression of the long (Ob-Rb) isoform of the leptin receptor in the brain, with the highest density being expressed in the hypothalamus [24]. The shorter leptin receptor isoforms are distributed throughout the body with the highest density being measured in the choroid plexus and leptomeninges, suggesting a role for leptin transport across the blood-brain barrier [23, 24]. As leptin is a large protein, it was proposed that it would be unable to cross the blood-brain barrier, from the blood to the cerebrospinal fluid where it is able to interact with leptin receptors in the brain. Therefore, a specific leptin transport system was proposed and subsequent experiments with leptin did in fact demonstrate an energy-dependent saturable transport system for leptin localised around the choroid plexus and median eminence, that could transport leptin from the blood into the CSF where it would have access to the hypothalamus.

The importance of the intracellular signal transduction component of the leptin receptor in mediating the actions of leptin was realised when it was discovered that the genetically obese mouse strain *db/db* was unable to produce the long-form (Ob-Rb) of the leptin receptor [25]. In *db/db* mice, a G to T mutation at precisely the junction where the long (Ob-Rb) and short form (Ob-Ra) of the leptin receptor transcripts diverge results in expression of only short forms of the leptin receptor in the brains of these mice. It was shown that leptin injections to *db/db* mice had no effects on body weight or food intake [20], demonstrating the importance of the intracellular component of the leptin receptor long-form in mediating leptin's signal transduction.

Further evidence of the role that the extracellular domain plays in the signal transduction mechanism of leptin receptors was discovered when it was found that the obese (*fa/fa*) Zucker rat has a mutation in a single base, resulting in a Glu^{269} to Pro^{269} amino acid substitution [26]. Although this mutation affects all the leptin receptor isoforms, and does not appear to alter the binding characteristics of leptin at the receptor, it is thought that the amino acid substitution significantly interferes with receptor dimerisation, which is thought necessary for activation of protein kinase activity. The effect of the *fa/fa* mutation is that the dose-response curve for leptin's ability to inhibit feeding is shifted by approximately one log unit to the right. This means that within normal physiological leptin concentrations, activity of the leptin receptor is significantly reduced, resulting in hyperphagia and reduced energy expenditure, similar to the phenotype of *ob/ob* and *db/db* mice.

Role of Leptin in Human Energy Balance

Soon after the discovery of leptin, it was suggested that obese humans might be unable to produce sufficient amounts of leptin to counter their increasing body fat mass. However, studies on obese subjects showed that most had higher plasma leptin concentrations than lean subjects, in direct proportion to their increased body fat mass [27]. This observation did not therefore agree with the absolute leptin deficiency hypothesis. A few individuals have, however, been identified with extreme or morbid obesity which can be attributed to a mutation in the *ob* gene such that they are unable to produce leptin, in much the same way as with *ob/ob* mice [28]. These individuals, who show early onset obesity, hyperphagia and reduced thermogenesis, confirm that leptin is essential for control of normal body weight in humans. Recently, a homozygous mutation in the leptin receptor gene was identified in three obese sisters resulting in a truncated leptin receptor protein which lacks both the intracellular and transmembrane domains [29]. These sisters exhibited early onset, extreme obesity that was characterised by hyperphagia and an interesting psychological profile which included irritability and a noted aggressiveness towards obtaining food. Furthermore, these individuals also had hypogonadotrophic hypogonadism, emphasizing the importance of intact leptin signalling for normal sexual maturation. These family members confirmed that in the absence of leptin signalling mechanisms in humans, there is a marked dysregulation of energy homeostasis and other neuroendocrine systems thought to be regulated by leptin in animal models. Studies of families in which siblings exhibit morbid obesity, like their parents, have provided some evidence for linkage of the human *ob* gene region with the development of the extreme obesity [30, 31]. However, it is clear that for a majority of moderate-to-severely obese humans the absence of significant linkage with the *ob* locus suggests that abnormal leptin synthesis is not the cause of the variation in body weight.

Leptin Interactions with the Hypothalamic NPY System

As leptin and NPY have opposing actions on energy balance, and given the clear abnormalities of hypothalamic NPY in leptin-deficient and leptin-resistant models, it was suggested that these systems may directly interact with one another such that decreased leptin activity may produce an increase in NPY activity. Conversely, increased leptin activity may then reduce activity of hypothalamic NPY. Leptin receptors have been localised to the NPY synthesising neurons in the ARC, the source of the majority of hypothalamic neuropeptide Y, which would support this contention [24]. The demonstration that genetically obese animals, *ob/ob* mice and *fa/fa* Zucker rats, which have reduced leptin signalling mechanisms, also have high hypothalamic NPY expression, as compared to lean controls, was evidence for a tonic inhibition of the NPY system by leptin and that the hyperphagia in these animals was NPY driven. Therefore, a reduction in leptin signalling, when fat stores are depleted during starvation or dietary restriction, results in a disinhibiton of NPYergic neuronal activity, leading to a doubling of NPY synthesis and increased NPY synaptic activity leading to a downregulation of NPY receptors. Injections of leptin to rats and mice has the opposing actions on the hypothalamic NPY system, with reductions in NPY synthesis [32, 33]. It also seems likely that the impaired reproductive function seen in *ob/ob* and *db/db* mice, and attributable to hypogonadotrophic hypogonadism is related to leptin deficiency, and the subsequent effects on NPY synthesis. In situations of negative energy balance, such as during food deprivation, exercise and lactation leptin levels fall, and there is a subsequent rise in NPYergic activity in the hypothalamus. This has the dual effect of increasing feeding behaviour, conserving energy by reducing thermogenesis, and switching off LH secretion and thus reproductive capacity (fig. 3).

The data presented so far suggest a simple feedback loop, with only two principal mediators, leptin and NPY; however there are several pieces of evidence which suggest that the system is more complex than this, and that several other (perhaps undiscovered) systems may be involved. The most compelling evidence comes from studies of NPY-knockout mice, and combinations of this mutation

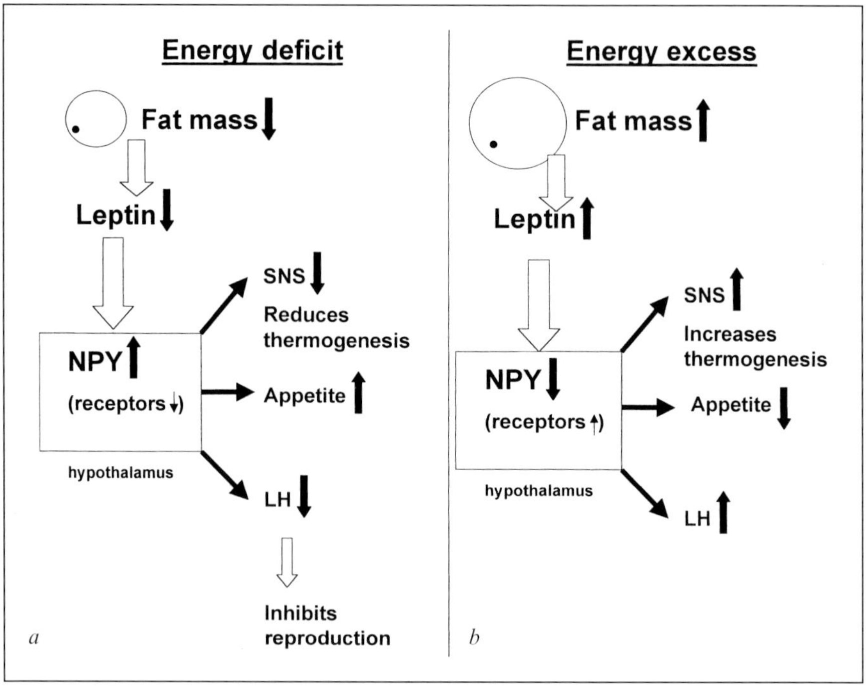

Fig. 3. Leptin-NPY interactions. *a* When there is an energy deficit, circulating leptin concentrations fall and hypothalamic NPY production and secretion increase, enhancing appetite, reducing thermogenesis and switching off LH secretion. *b* In situations of energy excess, the NPY system is turned off by the high leptin levels.

In genetic defects of leptin synthesis or with leptin receptor defects, NPY production remains high in the presence of energy excess. Reduced thermogenesis and increased appetite both contribute to the development of obesity. In dietary obesity NPY secretion is usually reduced, but the degree to which this occurs could alter the susceptibility of rats to obesity in this model.

with the *ob* gene mutation seen in obese *ob/ob* mice. The NPY-knockout mice are phenotypically normal, and show normal responses to food deprivation. They are, however, unusually sensitive to the anorexic effects of leptin, and are more prone to seizures than wild-type animals. Combination of the NPY knockout with the *ob* mutation produces an animal which, in terms of food intake, is midway between the *ob/ob* and lean phenotypes.

Is There Evidence for Leptin Resistance in Obesity?

Clearly, in genetic models of obesity, such as in the *db/db* mouse and *fa/fa* Zucker rat where leptin receptor activity is compromised, leptin action is significantly attenuated and these animals can therefore be considered to be leptin resistant. However, much less is known about nongenetically-induced obesity, such as the obesity resulting from the over consumption of highly palatable diets with a high fat content, which is a predominant causal factor leading to obesity in humans. Dietary-induced obesity in rodents more closely exhibits features seen in human obesity [14]. Allowing rats to overconsume a highly palatable diet with a higher fat content than normal pelleted chow results in obesity over a 6- to 8-week period. These obese rats exhibit a 100% increase in body fat pad mass and an approximate doubling in plasma leptin concentrations. Clearly, as with humans, the gradual rise in plasma leptin concentrations are not able to counter the rise in body weight, and it has been suggested that humans and rats that progress to obesity have become resistant to the effects of leptin. However, an examination of the hypothalamic NPY system in dietary-induced obese rats does not support this contention since elevations in plasma leptin produce, as expected, significant reductions in the activity of the hypothalamic NPY system. Furthermore, an examination of the density of hypothalamic NPY receptor density in dietary-obese rats reveals a receptor up-regulation, which is consistent with a reduction in NPY neuronal activity [14]. Clearly, the NPYergic neurones in the hypothalamic ARC are still able to sense the rising concentrations of leptin and respond with a reduction in their activity. If these neurones had become resistant to the actions of leptin, then one may expect normal or even increased activity of the NPYergic neurones Clearly then, endogenous leptin is still exerting a negative feedback on the hypothalamic NPY system suggesting that an overactive NPY system is unlikely to be responsible for the hyperphagia which leads to the development of obesity in this model. It is, however, possible that variations in the ability to suppress NPY in response to increased leptin concentrations might help explain why some strains of rat are resistant to dietary obesity, whilst others are sensitive. It remains to be seen whether further inhibition of this system, either by giving additional exogenous leptin, or by using selective NPY antagonists, will have any effect in this model of obesity.

Conclusions

The leptin-NPY system provides an important feedback control mechanism to regulate energy metabolism and ultimately body fat mass. During periods of negative energy balance, when there is a reduction in availability of food sources, or when there is increased energy disposal via increased physical exercise or thermogenesis, plasma leptin concentrations are reduced in proportion to decreases in body fat mass. The effect of a decrease in plasma leptin concentration is to cause a disinhibition of the hypothalamic NPY system, which is normally tonically inhibited by high leptin concentrations. The resultant increase in hypothalamic NPY activity stimulates appetite, to increase the calorific intake in response to the shift in negative energy balance and to conserve existing fat stores by reducing thermogenesis. Obesity in humans and rodents, resulting from the overconsumption of energy-rich diets, results in increased leptin secretion by adipose tissue, resulting in decreased activity of the hypothalamic NPY system. The regulation of the hypothalamic NPY system by leptin provides an important component in controlling body weight to within ideal limits. Dysregulation of the circuit, for example by mutations in the genes encoding leptin or the leptin receptor, results in the overactivation of the hypothalamic NPY system leading to overconsumption of food and obesity. In these circumstances, the hypothalamus fails to detect that there has been a shift towards positive energy balance, with the individual responding as if it were starving. It has yet to be conclusively determined whether increased caloric intake in a majority of obese humans can be attributed to a dysregulation of the leptin-NPY system or whether drugs which block the activity of NPY 'feeding' receptors will be therapeutically useful in obesity.

NPY is negatively regulated by leptin, which is transported into the brain across the blood-brain barrier, where it interacts with specific leptin receptors located on NPY containing neurones. The actions of NPY are to cause a shift in energy balance towards net energy gain through two mechanisms, an increase in food intake and a reduction in energy disposal via thermogenesis. Therefore, by decreasing NPY activity, leptin attenuates the activity of the NPY system, and, with the aid of other neurochemical systems, reduces food intake and increases thermogenesis.

References

1 Sawchenko PE, Swanson LW, Grzanna R, Howe PR, Bloom SR, Polak JM: Colocalization of neuropeptide Y immunoreactivity in brainstem catecholaminergic neurons that project to the paraventricular nucleus of the hypothalamus. J Comp Neurol 1985;241:138–153.

2 Zaijevski N, Cusin I, Rohner-Jeanrenaud F, Jeanrenaud B: Chronic intracerebroventricular neuropeptide Y administration to normal rats mimics hormonal and metabolic changes of obesity. Endocrinology 1993;133:1753–1758.

3 Egawa M, Yoshimatsu H, Bray GH: Neuropeptide Y suppresses sympathetic activity to interscapular brown adipose tissue in rats. Am J Physiol 1991;260:R328–R334.

4 Kalra SP, Dube MG, Sahu A, Phelps CP: Neuropeptide Y secretion increases in the paraventricular nucleus in association with increased appetite for food. Proc Natl Acad Sci USA 1991;88:10931–10935.

5 Lambert PD, Wilding JPH, Gilbey SG, Ghatei MA, Bloom SR: Effect of food deprivation and streptozotocin-induced diabetes on hypothalamic neuropeptide Y release as measured by a radioimmunoassay-linked microdialysis procedure. Brain Res 1994;656:135–140.

6 Lambert PD, Wilding JPH, Al-Dokhayel AAM, Bohuon C, Comoy E, Gilbey SG, Bloom SR: A role for neuropeptide Y, dynorphin and noradrenaline in the central control of food intake after food deprivation. Endocrinology 1993;133:29–32.
7 Akabayashi A, Wahlestedt C, Alexander JT, Leibowitz SF: Specific inhibition of endogenous neuropeptide Y synthesis in arcuate nucleus by antisense oligonucleotides suppresses feeding behavior and insulin secretion. Brain Res Mol Brain Res 1994;21:55–61.
8 Moltz JM, McDonald JK: Neuropeptide Y: Direct and indirect action on insulin secretion in the rat. Peptides 1985;6:1155–1159.
9 Kalra SP, Kaira PS: Nutritional infertility – The role of the interconnected hypothalamic neuropeptide Y-galanin-opioid network. Front Neuroendocrinol 1996;17:371–401.
10 Wilding JPH, Ajala MO, Lambert PD, Bloom SR: Additive effects of lactation and food restriction to increase hypothalamic neuropeptide Y mRNA in rats. J Endocrinol 1997;152:365–369.
11 Widdowson PS, Upton R, Henderson L, Buckingham R, Wilson S, Williams G: Reciprocal regional changes in brain NPY receptor density during dietary restriction and dietary-induced obesity in the rat. Brain Res 1997;774:1–10.
12 Stanley BG, Magdalin W, Seirafi A, Nguyen MM, Leibowitz SF: Evidence for neuropeptide Y mediation of eating produced by food deprivation and for a varient of the YI receptor mediating this peptide's effect. Peptides 1992;13:581–587.
13 Gerald C, Walker MW, Criscione L, Gustafson EL, Batzlhartmann C, Smith KB, Vaysse P, Durkin MM, Laz TM, Linemeyer DL, Schafthauser AO, Whitebread S, Hofbauer KG, Taber RI, Branchek TA, Weinshank RL: A receptor subtype involved in neuropeptide-Y-induced food-intake. Nature 1996;382:168–171.
14 Schafthauser AO, Stricker Krongrad A, Brunner L, Cumin F, Gerald C, Whitebread S, Criscione L, Hofbauer KG: Inhibition of food intake by neuropeptide Y Y5 receptor antisense oligonucleotides. Diabetes 1997;46:1792–1798.
15 Zhang Y, Proenca R, Maffei M, Barone M, Leopold L, Friedman JM: Positional cloning of the mouse *obese* gene and its human homologue. Nature 1994;372:425–432.
16 Maffei M, Halaas J, Ravussin E, Pratley RE, Lee GH, Zhang Y, Fei H, Kim S, Lallone R, Ranganathan S, Kern PA, Friedman JM: Leptin levels in human and rodent: Measurement of plasma leptin and ob ma in obese and weight-reduced subjects. Nat Med 1995;1:1155–1161.
17 Halaas IL, Gajiwala KS, Maffei M, Cohen SL, Chait BT, Rabinowitz D, Lallone RL, Burley SK, Friedman JM: Weight-reducing effects of the plasma protein encoded by the obese gene. Science 1995;269:543–546.
18 Campfield LA, Smith FJ, Guisez Y, Devos R, Burn P: Recombinant mouse ob protein: Evidence for a peripheral signal linking adiposity and central neural networks. Science 1995;269:546–549.
19 Stephens TW, Basinski M, Bristow PK, Bue-Vallesky JM, Burgett SJ, Crafi L, Hale J, Hofmann J, Hsuing UM, Kriauciunas A, et al: The role of neuropeptide Y in the antiobesity action of the obese gene product. Nature 1995;377:530–532.
20 Tartaglia LA, Dembski M, Weng X, Deng NH, Culpepper J, Devos R, Richards GJ, Campfield LA Clark FT, Deeds J, Muir C, Sanker S, Moriarty A, Moore KJ, Smutko JS, Mays GG, Woolf EA, Monroe CA, Tepper RI: Identification and expression cloning of a leptin receptor, ob-r. Cell 1995;83:1263–1271.
21 Mercer JG, Hoggard N, Lawrence CB, Rayner DV, Trayhurn P: Localization of leptin receptor (ob-r) gene-expression in mouse-brain by in-situ hybridization. J Physiol Lond 1996;495P:113.
22 Chua SC, Chung WK, Wupeng XS, Zhang YY, Liu SM, Tartaglia L, Leibel RL: Phenotypes of mouse diabetes and rat fatty due to mutations in the OB (leptin) receptor. Science 1996;271:994–996.
23 Chua SC, White DW, Wupeng XS, Liu SM, Okada N, Kershaw EE, Chung WK, Powerkehoe L, Chua M, Tartaglia LA, Leibel RL: Phenotype of fatty due to gln269pro mutation in the leptin receptor (lepr). Diabetes 1996;45:1141–1143.
24 Considine RV, Sinha MK, Heiman ML, Kriauciunas A, Stephens TW, Nyce MR, Ohannesian JP, Marco CC, Mckee LJ, Bauer TL, Caro JF: Serum immunoreactive leptin concentrations in normal-weight and obese humans. N Engl J Med 1996;334:292–295.

25 Montague CT, Farooqi IS, Whitehead JP, Soos MA, Rau H, Wareham NJ, Sewter CP, Digby JE, Mohammed SN, Hurst JA, Cheetham CH, Earley AR, Barnett AH, Prins JB, O'Rahilly S: Congenital leptin deficiency is associated with severe early-onset obesity in humans. Nature 1997;387:903–908.
26 Clement K, Vaisse C, Lahlou N, Cabrol S, Pelloux V, Cassuto D, Gourmelen M, Dma C, Chambaz J, Lacorte J, Basdeveant A, Bougneres P, Lebouc Y, Froguel P, Guy-Grand B: A mutation in the leptin receptor gene causes obesity and pituitary dysfunction. Nature 1998;392:398–401.
27 Clement K, Garner C, Hager J, Philippi A, Leduc C, Carey A, Harris TJR, Jury C, Cardon LR, Basdevant A, Demenais F, Guygrand B, North M, Froguel P: Indication for linkage of the human ob gene region with extreme obesity. Diabetes 1996;45:687–690.
28 Reed DR, Ding Y, Xu WZ, Cather C, Green ED, Price RA: Extreme obesity may be linked to markers flanking the human ob gene. Diabetes 1996;45:691–694
29 Schwartz MW, Seeley RI, Campfield LA, Burn P, Baskin DG: Identification of targets of leptin action in rat hypothalamus. J Clin Invest 1996;98:1101–1106.
30 Wang Q, Bing C, Al-Barazanji K, Mossakowaska DE, Wang X, McBay D, Neville WA, Taddayon M, Pickavance L, Dryden S, Thomas MEA, McHale MT, Gloyer IS, Wilson S, Buckingham R, Arch JRS, Trayhurn P, Williams G: Interactions between leptin and hypothalamic neuropeptide Y neurones in the control of food intake and energy homeostasis in the rat. Diabetes 1997;46:335–34

J.P.H. Wilding, DM, MRCP, Diabetes and Endocrinology Research Unit, Department of Medicine, Liverpool University, Duncan Building, Daulby Street, Liverpool L69 3GA (UK)
Tel. +44 151 706 4070, Fax +44 151 706 5797, E-Mail J.P.H.Wilding@liverpool.ac.uk

Ur E (ed): Neuroendocrinology of Leptin.
Front Horm Res. Basel, Karger, 2000, vol 26, pp 87–105

Perspectives on Leptin's Role as a Metabolic Signal for the Onset of Puberty

Clement C. Cheung[a], *Donald K. Clifton*[b], *Robert A. Steiner*[a–c]

Department of [a] Physiology and Biophysics, [b] Obstetrics and Gynecology, and [c] Zoology, University of Washington, Seattle, Wash., USA

In higher mammals, the brain governs the timing of sexual maturation. Under normal physiological conditions, the orchestration of pubertal development resides in a network of gonadotropin-releasing hormone (GnRH) neurons in the forebrain. The importance of GnRH neurons in sexual maturation is illustrated by clinical observations revealing that several forms of nonidiopathic precocious puberty are GnRH-dependent [1]. In addition, individuals with Kallmann's syndrome have abnormal migration of their GnRH neurons during intrauterine development, which becomes manifest as hypogonadotropic hypogonadism and a failure to undergo normal sexual development at puberty. Other clinical observations reveal that abnormalities in brain function, resulting from infection, trauma, irradiation, congenital malformations, and hypothalamic hamartoma are linked to the advancement or delay of pubertal maturation [1]. GnRH neurons become activated briefly during the neonatal period, return to quiescence throughout the juvenile period, and become reactivated just prior to the onset of puberty [2]. Among the proposed mechanisms governing the activation of GnRH neurons at the time of puberty are those involving intrinsic alterations in the activity of certain key neurotransmitters and neuromodulators. Some of these inputs are postulated to exert a restraint on the secretory capability of GnRH neurons during the prepubertal period, which then diminish to allow puberty to progress, whereas other inputs are theorized to impart an excitatory input to activate the GnRH system at puberty [2]. The cellular and molecular mechanisms triggering the shift in the balance of these putative inhibitory and excitatory inputs to the GnRH system remain unknown. In addition, factors outside of the central nervous system have been implicated in the initiation of puberty. Among these are nutritional and metabolic signals, which have been implicated in the initiation and maintenance of GnRH secretion [3].

Nutrition and Puberty

Clinical observations and research on laboratory animals and domesticated species have demonstrated a strong correlation between availability of nutrients and sexual maturation [4, 5]. In rats, underfeeding retards growth and delays the onset of puberty. This fuel-limited restraint on sexual maturation can be rapidly reversed upon refeeding [6–9]. Similarly, studies in humans have shown that young girls who exercise heavily have delayed menarche, and this exercise-dependent interruption in pubertal development can be reversed with either sufficient weight gain or decreased athletic activity [10]. The effects of metabolic insufficiency on puberty are mediated by the brain and pituitary. Numerous studies have shown that food restriction inhibits the secretion of GnRH, luteinizing hormone (LH), and follicle-stimulating hormone (FSH) and that refeeding reverses this inhibition relatively quickly – within hours [7, 11].

Kennedy and Mitra [12] were the first to demonstrate that in the laboratory rat, body weight is more strongly correlated with the timing of puberty onset than chronological age. Frisch and Revelle, using the rat and human as their references, suggested that achieving a critical body weight somehow initiates sexual maturation in both of these species [13, 14]. Their theory was revised by others who proposed that a critical body composition predicts more accurately when puberty occurs [15]. For example, Wilen and Naftolin [5] suggested that the percentage of protein in body composition is important during pubertal development, whereas Frisch [16] focused on the notion that achieving a particular lean/fat ratio is a major factor in determining the onset of puberty.

Whatever their particular formulation, these theories were based on the observed close association between body composition or nutritional status and puberty. Implicit in this association is a means by which metabolic information can be conveyed to the reproductive axis – in particular to the brain; however, the identities of the molecular messengers serving this signaling function remain unknown. Over the years, several candidates have been advanced as possible metabolic signals to the reproductive system – insulin, glucose, and amino acids, among others [17–19]. There is physiological and anatomical evidence to support the argument that these hormones and fuel metabolites could be metabolic relays to the reproductive axis; however, none of them appears to act as the primary linkage between nutrition and sexual maturation. Leptin, a recently discovered hormone, has been added to the list of potential metabolic messengers and is being examined for its potential role as a link between metabolism, body composition, and reproductive function [20].

Leptin and Reproduction

Leptin is a secretory product of the *obese* (*ob*) gene that was first identified in 1994 [21]. Leptin is secreted from adipocytes (and the placenta and the stomach as reported recently) [21–25], and its circulating concentration is tightly correlated with body fat content and body mass index, with circulating levels of leptin being higher in the obese than in the lean state [26]. Plasma levels of leptin increase with weight gain and decrease with weight loss [26–28]. In addition, transient changes in energy intake, without any appreciable changes in body weight, can alter circulating leptin levels [28–32]. The expression of *ob* mRNA in adipocytes, like circulating leptin concentrations, is also affected by changes in weight and energy intake [33, 34]. The critical importance of leptin for body weight regulation and metabolism is best illustrated by the *ob/ob* mouse, a genetically obese mouse model. *ob/ob* mice were first discovered in the 1950s and are now recognized to be leptin-deficient due to either a nonsense mutation or a deletion in the *ob* gene [21, 35]. The myriad of metabolic abnormalities in *ob/ob* mice includes hyperphagia, hyperinsulinemia, hypometabolism, and progressive obesity [35, 36]. Not surprisingly, exogenous administration of leptin to these mice can decrease their food intake and normalize their body weight to approximate that of the lean, wild-type animal [37–39]. Rodents that harbor mutations in the leptin receptor-encoding gene, e.g. the *diabetic* mouse and the *fatty* rat, exhibit a similar phenotype; however, these animals manifest a reproductive phenotype that is less profoundly disturbed than the *ob/ob* mouse, suggesting that some leptin-dependent signaling may take place even with highly dysfunctional receptors being present [40–42]. Together, these observations suggest that leptin serves as a signal in a negative feedback loop that physiologically couples appetite, metabolism, and adiposity.

In addition to the metabolic abnormalities found in *ob/ob* mice, these animals also fail to achieve normal puberty, and both sexes are infertile [35]. Closer examination of the reproductive system of these animals reveals that the reproductive tracts of both sexes are profoundly atrophied. Their underdeveloped gonads show arrested gametogenesis, and circulating levels of gonadotropins and sex steroids are low compared to age-matched lean wild-type mice [35, 43–45]. These observations provide clear evidence that leptin is obligatory for fertility and suggest that leptin serves as a physiological link between metabolism and reproduction. Experimental studies with recombinant leptin reinforce the concept that leptin is required to support normal reproductive function. We and others have shown that reproductive dysfunction in *ob/ob* mice can be corrected with leptin replacement [46–48]. Our studies have demonstrated that injections of leptin in *ob/ob* mice increase plasma levels of gonadotropins and sex steroids and stimulate gonadal activity in both sexes [46]. In addition, studies by Chehab et al. [47, 48] have shown that leptin administration can correct the infertility of both male and female *ob/ob* mice. These observations demonstrate that leptin has an inductive effect on the reproductive axis of the *ob/ob* mouse.

That leptin is a metabolic signal to the reproductive system is further evidenced by the work of Ahima et al. [49], who have shown that in mice, suppression of the gonadal axis resulting from a 48-hour fast can be prevented by leptin treatment. In their experiments, a 2-day fast resulted in a decrease in levels of LH and testosterone in males and a delay in vaginal estrus in females. As expected, serum leptin levels fell during the 48-hour fast; however, mice supplemented with leptin during the fast had indices of reproductive function that were indistinguishable from those of normal mice. Leptin's ability to reverse the suppressive effect of fasting on the reproductive system has been extended to nonhuman primates as well. In a study designed to test the hypothesis that leptin is a metabolic activator of the reproductive axis in the male rhesus macaques, Finn et al. [50] administered leptin to fasting monkeys and compared their reproductive hormone levels to other fasting monkeys that had received saline only. Withholding food for 48 h completely abolished the pulsatile release of LH in saline-treated monkeys during the second day of the fast. In contrast, monkeys that received peripheral leptin infusion during the fast had significantly higher mean FSH and LH levels, as well as greater LH pulse frequency and amplitude. The observation that exogenous leptin can restore reproductive function in animals that are deficient in leptin argue that a foundation of circulating leptin is necessary to sustain normal reproduction and that plasma levels of leptin below a critical threshold become manifest in reproductive failure. As a corollary, we infer that leptin acts as a signal that gates the activity of the reproductive system in response to changes in the status of fat reserves and metabolism.

Leptin and Puberty in the Rat

Leptin's close association with reproduction implicate this hormone as a metabolic signal for the onset of puberty. *ob/ob* mice, which lack leptin, are delayed in their pubertal development, suggesting leptin may be involved in

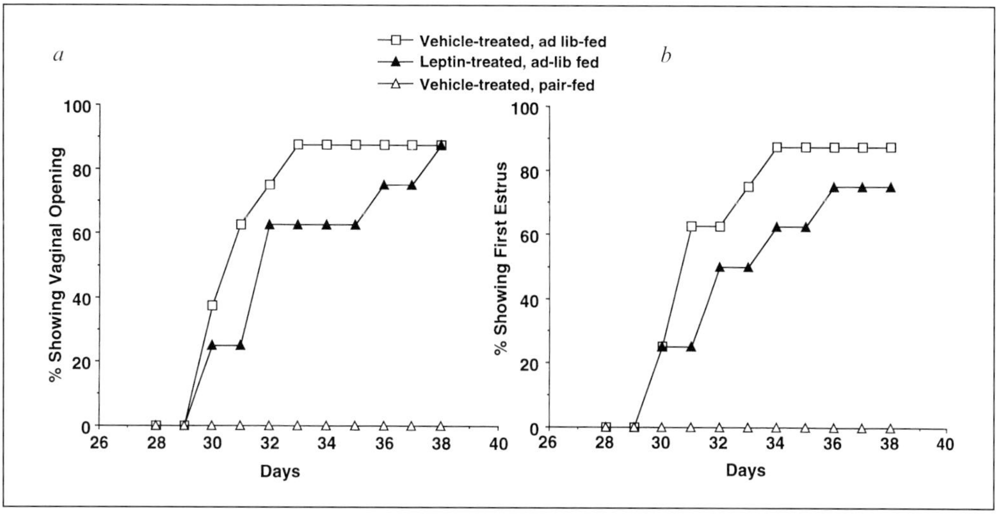

Fig. 1. Percentage of animals showing (*a*) vaginal opening and (*b*) first estrus. Animals were divided into three groups: the first group received vehicle injection and were fed ad libitum (open square), the second group received 6.3 μg/g twice daily injection of leptin and were fed ad libitum (filled triangle), and the third group received vehicle injection and were pair-fed with the leptin-treated animals (open triangle).

the normal timing for the onset of puberty [35]. We hypothesized that if leptin were the primary signal for initiating the onset of puberty, then a premature increase in circulating leptin, at least in the female rat, would initiate an earlier onset of puberty – reflected in the mean age of vaginal opening and first estrus. We tested this hypothesis by administering recombinant human leptin (6.3 μg/g, i.p., twice daily) to prepubertal female Sprague-Dawley rats starting on day 23 of life, 2 days after weaning [51]. Two other groups received vehicle injections and were either fed ad libitum or pair-fed to the leptin-treated group. (Pair-feeding controls for the inhibitory effects exerted on the reproductive axis due to a decrease in food intake, a major consequence of leptin treatment.) Animals were sacrificed on the first day of diestrus, 2 days after they had undergone their first ovulatory surge. Prepubertal rats that had been treated with leptin showed a food consumption at 80% of normal, together with a drop in weight when compared to those that were fed ad libitum. Notwithstanding this food reduction and weight loss, the rate of sexual maturation in these leptin-treated rats, as reflected by their mean age of vaginal opening and first estrus, was indistinguishable from vehicle-treated, ad-libitum-fed animals; whereas in the pair-fed group, the same measures of sexual maturation were significantly delayed (fig. 1). Histological analyses showed that by 38 days of

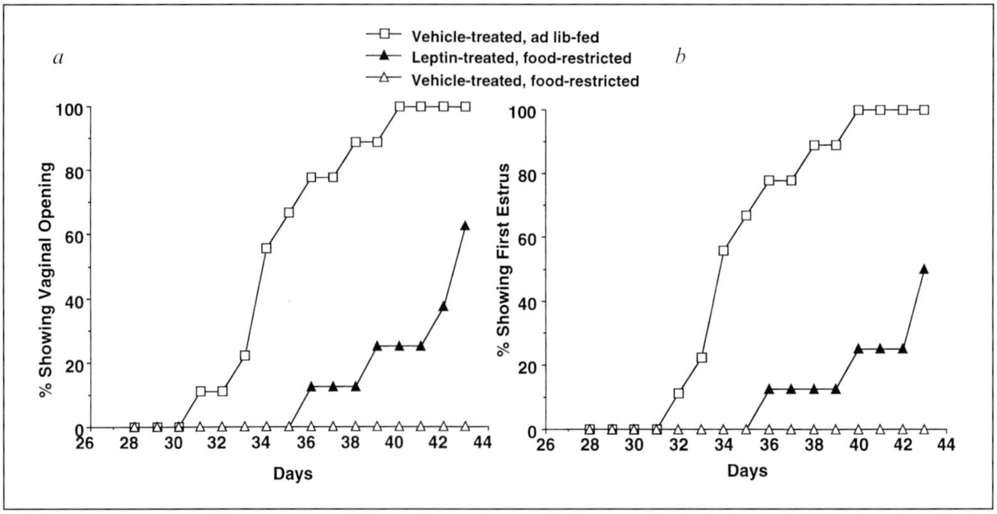

Fig. 2. Percentage of animals showing (*a*) vaginal opening and (*b*) first estrus. Animals were divided into three groups: the first group received vehicle injection and were fed ad libitum (open square), the second group received 6.3 μg/g twice daily injection of leptin and were food restricted to 70% of normal (filled triangle), and the third group received vehicle injection and were also food-restricted to 70% of normal (open triangle).

age, corpora lutea were present in 7 of 8 animals in both the ad-libitum-fed groups, thus confirming ovulation; however, corpora lutea were completely absent in the pair-fed animals. Reproductive organ weights were significantly greater in the leptin-treated and vehicle-treated, ad-libitum-fed animals compared to those that were pair-fed. These observations establish that in rats, the delay in sexual maturation caused by a 20% reduction in food intake and a decrease in body weight can be prevented by leptin treatment.

To further our understanding of the role of leptin in influencing the timing of pubertal onset in the rat, we imposed a more severe food restriction during development (70% of normal intake) and assessed the effect of leptin on the rate of sexual maturation [51]. Prepubertal female rats were divided into three groups – the first group received vehicle injections and were fed ad libitum. The other two groups were fed at 70% of the first group; one of these two received leptin (6.3 μg/g, i.p., twice daily), whereas the other received vehicle injections only. By 43 days of age, 5 of the 8 leptin-treated, food-restricted rats had shown vaginal opening, and 4 of those had attained first estrus. On the contrary, none of the vehicle-treated, food-restricted female rats had shown any signs of vaginal opening by 43 days of age (fig. 2), suggesting that they were significantly delayed in their rate of sexual maturation – despite having had the same level of food restriction as the leptin-treated animals. However, unlike the results of the first experiment in which leptin-treated animals had a rate of sexual maturation comparable to normal animals despite a 20% decrease in food intake, leptin administration to animals that were 30% food-restricted did not achieve a rate of sexual maturation that was equivalent to that of the ad-libitum-fed animals.

Our observation that the delay in the onset of puberty associated with food restriction can be partially reversed by leptin is corroborated by the work of Gruaz et al. [52] who showed additionally that in 25-day-old prepubertal female rats that were severely food-restricted (36% of normal intake), serum leptin levels were drastically reduced compared to normal developing rats and that this reduction is associated with a complete failure of pubertal development. When these food-restricted animals reached 53 days of age, they were given exogenous leptin administered into the cerebral ventricles (i.c.v.; 10 μg/day), which produced a rate of vaginal opening comparable to food-restricted animals of the same age that were allowed to feed ad libitum. These results suggest that the delay in sexual maturation caused by food restriction occurs as a direct result of a decline in circulating leptin to levels below those necessary for activating the reproductive axis at the normal time of puberty. However, if this threshold for activation in these food-restricted rats can be subsequently achieved by either refeeding or artificially raising the concentration of leptin in the brain, the normal progression of pubertal maturation can be resumed.

Although leptin is critical for the onset of puberty in mice, as reflected by the arrested pubertal onset in *ob/ob* mice, our experiments in rats also demonstrate that treatment of prepubertal animals with high levels of leptin does not advance the normal timing of pubertal onset. However, it does sustain the normal progression of pubertal development – despite a moderate decrease (a 20% reduction in our study) in food intake that would normally (i.e. without leptin treatment) have induced a delay in sexual maturation [51]. Peripherally administered leptin is also successful, albeit only partially, in counteracting the inhibitory effect of a severe food restriction (a 30% reduction in our study) on the onset of puberty. In the experiments of Gruaz et al. [52], the suppression in pubertal development caused by an even more severe dietary restriction (36% of the normal intake) can be reversed by central administration of leptin or refeeding, with equal effectiveness in both groups of animals. Together, these results argue that leptin is not the rate-determining factor in the timing of puberty onset, but rather that leptin acts as a permissive factor to gate the onset of puberty as a function of the animal's metabolic state. Thus, on the one hand, sexual maturation will proceed normally when nutritional status is adequate, reflected by plasma leptin levels that reach or exceed a certain threshold; on the other hand, sexual maturation will be delayed if nutrition and fat reserves are inadequate, reflected by plasma leptin levels that fall below this critical threshold.

The proposal that leptin exerts a permissive effect on puberty is supported by our observation that, in the female rat, serum leptin levels remain relatively steady from day 10 to day 50, across the normal span of pubertal development [unpubl.]. Leptin levels begin to rise only after animals have entered the young adult stage at approximately day 50. The relatively stable circulating concentrations of leptin across the boundary between prepubertal life and sexual maturation would suggest that although a certain plasma threshold of leptin is requisite for sexual development, achieving this threshold is insufficient to activate the reproductive axis before the event would normally occur (since this same level is present as early as 10 days after birth). Clearly, other factors besides leptin are necessary to fully initiate sexual maturation in the female rat. Therefore, increasing plasma levels of leptin per se are not the triggering event for initiating puberty. Nevertheless, leptin levels falling below this threshold (e.g. in the *ob/ob* mouse or with food restriction in any species) impose a severe brake on the normal timing of pubertal onset. Although leptin levels were not measured in our earlier studies, we can infer from the work of others, in which fasting was reported to reduce leptin levels, that serum leptin concentrations in our food-restricted female rats were indeed lower than those in normal ad-libitum-fed animals [28, 29, 31].

Although leptin originates in the periphery, it crosses the blood-brain barrier by a saturable transport process [53, 54]. Leptin's target sites in the brain are many, as suggested by the wide distribution of leptin receptor mRNA [55–57]. The leptin receptor isoform that possesses signaling ability has been identified in areas of the brain that are known to be involved in feeding and reproduction, including the arcuate, dorsomedial, and ventromedial nuclei of the hypothalamus [57, 58]. Therefore, it seems highly plausible that leptin mediates its effects on sexual development by acting directly on the central nervous system. The work from Grauz et al. [52] shows that leptin given centrally can reverse the effect of food restriction on sexual maturation. The inference that leptin can mediate its effect on the reproductive axis by acting on targets in the brain is supported by the observation that at the dosage used in the central infusion study, peripheral leptin concentration remains unchanged. Recently, work by Nagatani et al. [59] has provided additional evidence for leptin's central effect on the reproductive endocrine system. Their experiments have revealed that when ovariectomized female rats (treated with vehicle or estrogen) are subjected to a 48-hour fast, the subsequent suppression in pulsatile LH secretion can be reversed by leptin given by an intraperitoneal route. Specifically, the attenuation of LH pulse frequency and amplitude during fasting are normalized by leptin administration. These findings suggest that leptin stimulates LH secretion, possibly through a direct effect on the brain-pituitary axis. Although the pituitary cannot be ruled out as a target for the action of leptin [60], the increase in LH pulse frequency suggests an interaction between the GnRH pulse generator in the hypothalamus and leptin. However, this interaction is likely to be indirect, because leptin receptor expression in the preoptic area, where GnRH neurons are located in the rat, is scarce. Thus, it seems likely that leptin acts indirectly on the GnRH pulse generator through other systems that modulate GnRH secretory activity in the hypothalamus.

Strengthening the argument that leptin is an important factor, but perhaps not *the* triggering factor, in controlling the timing of puberty are studies that examine the effects of leptin administration on sexual maturation in other species. In addition, studies correlating circulating leptin levels and the timing of the onset of puberty have been performed in the rat, mouse, monkey and human. Although common themes have been observed among these species, there are also notable differences among them that suggest caution in drawing conclusions about leptin's possible role in sexual maturation across species.

Leptin and Puberty in the Mouse

Although studies performed in the female rat suggest that leptin may act as a permissive signal to the onset of puberty, similar studies performed in the mouse paint a somewhat different and more controversial picture. Chehab et al. [47] showed that vaginal opening can be advanced in female C57BL/6J mice that were injected daily with 2 μg/g leptin starting at 21 days of age. Four days after the start of injections, 100% of the leptin-treated mice had shown vaginal opening, whereas only 25% of those treated with phosphate-buffered saline (PBS) had shown vaginal opening. The onset of puberty occurred earlier in these leptin-treated mice, despite a decrease in their body weight and food intake. In addition, copulatory plugs were detected at an earlier age in those treated with leptin, again suggesting that the acceleration in sexual maturation, induced by leptin, encompasses both the external morphology and the internal circuitry that enables mating and successful pregnancy to occur. However, a second study carried out by the same group of investigators under an identical experimental paradigm revealed no apparent difference in the date of vaginal opening between leptin-treated and PBS-treated mice [48]. A similar finding was also observed by our own laboratory where leptin injection in prepubertal female C57BL/6J mice starting at 21 days of age did not alter the average date of either vaginal opening or first estrus [unpubl.]. However, the date of first estrus was advanced in leptin-treated mice compared to vehicle-treated mice which were pair-fed to the leptin-treated group. Therefore, our results in the mouse corroborate the data from our earlier studies in the female rat, suggesting that leptin may indeed act as a permissive signal to the onset of puberty [51].

In a similar study conducted by Ahima et al. [61], female C57BL mice were injected with either 2 μg/g leptin or vehicle starting at 21 days of age. Both groups of prepubertal mice showed similar weight gain throughout the study yet the onset of sexual maturation was earlier in the leptin-treated animals. Average date of vaginal opening, first estrus, and vaginal cycling were significantly advanced in leptin-treated compared to saline-treated mice. The difference in the mean age of vaginal opening between the two groups was approximately 1 day, and the difference in the mean age of first estrus and cycling was 3 and 2 days, respectively. Ahima et al. [61] further showed in normal mice that increasing body weight and estradiol levels across development are accompanied by an unchanging level of serum leptin in the prepubertal stage. Although exogenous leptin administration may have initiated the acceleration in the aforementioned pubertal indices, the physiological significance of this advancement is challenged by the observation that endogenous leptin levels remained stable – despite an increase in body weight and circulating levels of estradiol throughout the pre- and postpubertal period. In fact, levels of endogenous leptin measured in animals on their day of vaginal opening did not differ from those of animals at the same age whose vaginas remained closed. The absence of a rise in leptin levels prior to the appearance of these pubertal indices argues against the supposition that achieving a threshold of circulating leptin levels acts as a trigger for the onset of puberty. Rather, steady-state leptin levels across puberty seem to support the idea that achieving a threshold leptin level is an essential but insufficient cue to activate the reproductive axis at puberty. Leptin as a metabolic signal becomes important to the onset of puberty if its level deviates negatively from this critical threshold level, as is the case with either food restriction or excessive use of metabolic reserves (as in chronic disease or heavy exercise); under these circumstances, normal progression of pubertal development will be delayed or blocked entirely.

Interestingly, serum leptin levels in the mouse and the rat are reportedly elevated during the early preweaning stage and subsequently decline to a stable level throughout the peripubertal period [52, 62–64]. In addition, the surge in leptin levels during this early developmental period is apparently insensitive to food deprivation [64]. It has been postulated that this early surge in circulating levels of leptin may be important for the development of various neuroendocrine systems [64]. Whether the dissociation between plasma leptin levels and body weight during early development is important for the onset of puberty remains unanswered. The inconsistency in the various reports of leptin's effects on the timing of the onset of puberty in rodents and the apparent lack of a rising plasma titer of leptin just prior to puberty suggests that conclusions about leptin's possible role in controlling the timing of puberty onset should be treated with considerable caution. Species differences may account for some of these discrepancies. In addition, the observed differences in leptin's effect on body weight reported among the various mouse studies suggest that the differing sources of recombinant leptin used in these studies is another uncontrolled variable that could help to explain the disparate results.

Leptin and Puberty in the Monkey and Human

The idea that leptin is a permissive signal rather than a trigger for the onset of puberty is further supported by the developmental profile of serum leptin in the male rhesus monkey. Leptin levels in intact male monkeys remain stable across the time when puberty occurs [65, 66]. In castrated monkeys, the prepubertal rise in LH and FSH, which reflects the reactivation of GnRH neurons, is not preceded by any increase in serum leptin levels [65]. This apparent dissociation between the activation of the reproductive endocrine system and leptin levels lends support to the assertion that leptin does not directly determine the timing of pubertal onset, but rather acts as a metabolic 'gate' to the onset of puberty, allowing puberty to progress – provided that nutrition and metabolic reserves are sufficient, reflected by circulating levels of leptin above the lower-limit of the threshold value.

Developmental changes in serum leptin levels during puberty in the human are more varied and complex. Many studies have examined the relationship between serum leptin concentrations and the various stages of puberty, but the overall results are inconclusive. In addition, sexual dimorphism in leptin concentrations across puberty in light of the changing steroidal environment complicates any generalization made about leptin and puberty in our species [67–72]. In general, obese children have higher leptin levels than do lean children and the correlation between body weight and leptin levels is stronger in girls than in boys [67, 71, 73, 74]. Several cross-sectional studies in boys and girls have shown that prepubertal and postpubertal leptin levels do not differ significantly from each other [74, 75]. However, other reports testify to gender-specific differences in leptin levels across puberty. For instance, leptin levels in girls increase gradually with pubertal development and may serve as a signal for the onset of menarche [69, 73, 76, 77]. The rise in leptin levels may be related to the concomitant change in plasma estrogen levels during puberty. In support of this argument is the observation that estradiol may stimulate either leptin gene expression or increase circulating leptin concentrations [78, 79]; however, reports of this effect are equivocal [59].

In boys, both longitudinal and cross-sectional studies have shown that leptin levels increase initially and then drop after Tanner stage II or III [68, 69, 72, 76, 80, 81]. The rise in leptin levels precedes the pubertal increase in reproductive hormones, such as testosterone [68, 80]. Whether changes in leptin and testosterone are purely coincidental or causally related remain uncertain. It is conceivable that a rise in leptin levels may trigger the secretion of testosterone by acting directly on the testis; alternatively, leptin may increase testosterone secretion indirectly by stimulating the neuroendocrine reproductive axis. In addition, the fall in leptin levels during later Tanner stages may be a result of the suppressive effect of rising testosterone. This assertion is supported by several observations. First, leptin levels are lower in males than females, even after correction for body fat content [69]. Second, in boys with precocious puberty, leptin levels are increased by GnRH analogues, which block LH and testosterone secretion [82]. Third, in hypogonadal men, elevated serum leptin levels can be reduced by testosterone replacement [81], and fourth, testosterone inhibits leptin secretion and mRNA expression in cultured adipocytes [70]. Together, these observations suggest that there may be important interactions between sex steroids and leptin.

The importance of leptin to sexual development in humans can be further appreciated from a recent report by Strobel et al. [83] in which three members in a Turkish family were identified that have a homozygous mutation in their leptin gene. As expected, these three patients are obese and exhibit characteristics that are phenotypic features of *ob/ob* mice – hyperphagia, hyperinsulinemia, and hypothermia. In addition, the two adults in the cohort, one male and one female, exhibited hypogonadism.The female patient had primary amenorrhea while the male patient remained prepubertal at age 22. This male patient lacked secondary male sexual characteristics – sparse facial hair, bilateral gynecomastia, and underdeveloped testes and external genitalia. His serum total testosterone, free testosterone, and sex-hormone-binding globulin levels were below normal but could be simulated by the administration of human chorionic gonadotropin administration. The hypogonadotropic hypogonadism in this male patient is likely of hypothalamic origin, as a GnRH challenge test given to him normalized his LH and FSH levels. Despite the small sample size, the obvious linkage between a mutation in the leptin gene and abnormal reproductive characteristics suggests that leptin is critically important to human sexual development and reproductive capability.

Careful longitudinal studies on the leptin-deficient child in this cohort and two other children reported elsewhere should reveal valuable information as to what role leptin plays during the onset of puberty in human [83, 84]. If the clinical symptoms seen in these patients were entirely the result of leptin deficiency, then these patients may truly mimic the *ob/ob* mouse, and can potentially be treated by leptin replacement therapy. Under those circumstances, longitudinal information on leptin and puberty in the human may be unobtainable if the patient's reproductive states are corrected prior to puberty. To fully appreciate the effects of leptin deficiency on human pubertal development, we must rely on longitudinal studies of patients who carry a mutation in the leptin receptor gene [85] or retrospective studies of any as-yet-undiscovered adult patients who are homozygous recessive for the leptin gene.

Conclusions and Future Directions

This review summarizes various laboratory and clinical studies that have explored the possible role of leptin as a metabolic signal to the onset of puberty in mammalian species. While the state of leptin deficiency, which is reflected by the *ob/ob* mouse and patients with congenital leptin deficiency, reveals the critical importance of leptin on sexual maturation, questions remain as to *what* role leptin plays during pubertal development and *how* leptin exerts its effect on the reproductive axis.

According to published studies in the female mouse, leptin can accelerate the pace of sexual maturation when it is administered exogenously, with the inference that leptin acts as a triggering signal to the onset of puberty. However, firm conclusions about these observations are clouded by varius discrepancies in the methods and results of these studies. Results from studies in the rat and the monkey conclude that leptin does not act as a trigger but is instead a permissive signal for the onset of puberty – that is, some threshold level of plasma leptin is essential but alone is insufficient to allow normal sexual maturation to proceed [51, 52, 65, 66]. In the rat, pharmacological elevation of plasma leptin levels that exceed this threshold does not advance puberty; thus, in this species, leptin alone cannot be the rate-determining factor for the onset of puberty. It may be that leptin's role in puberty differs among species; this needs to be explored more fully. In addition, the small number of in vivo studies that have been performed (with mostly female animals) and the sexual dimorphism seen in human studies reinforce the need to revisit the conclusions of the original reports and to consider including both sexes in any further examination of leptin's possible role in puberty.

Aside from the active research into leptin's role in the physiology of reproduction, studies that focus on understanding the interaction between leptin and the central nervous system are also being persued intensively. It seems likely that leptin mediates its effect on puberty by interacting with various neuropeptides in the hypothalamus. Neuropeptide Y (NPY) and proopiomelanocortin (POMC) are good candidates as the peptide-containing neurons in the arcuate nucleus of the hypothalamus both coexpress leptin receptor mRNA and protein [86–88]. In particular, NPY has been proposed to play a significant role by being facilitatory as well as inhibitory to the onset of puberty in the female rat [89]. Persistent elevation of NPY in the arcuate nucleus, as seen in food restriction or chronic administration of NPY into the lateral ventricle, can significantly delay puberty [90, 91]. Since leptin inhibits NPY expression in the arcuate nucleus, a potential mechanism for leptin to modulate the effects of food restriction on sexual maturation is by diminishing the inhibitory tone that impinges on the reproductive axis through the reduction of arcuate NPY levels [92, 93]. In support of this hypothesis is the observation that food-restricted prepubertal rats attain sexual maturation upon refeeding but not refeeding coupled with i.c.v. administration of NPY [90]. However, NPY is not essential to leptin's effect on puberty, and other neuropeptides in the hypothalamus may mediate leptin's effect on sexual maturation in a manner similar to that of NPY [94]. POMC peptide products, such as α-melanocyte-stimulating hormone (α-MSH), may also participate in sexual maturation. Leptin stimulates POMC expression, and α-MSH has been reported to stimulate sexual maturation [93, 95–97]. By inference, α-MSH may mediate leptin's stimulatory effect on the onset of puberty, but whether NPY, POMC-associated peptides, or any other hypothalamic neuropeptides are important for leptin's effect on puberty remains unknown. A wealth of information will no doubt be revealed in the coming years about the relationship between leptin and sexual maturation. Until then, understanding the mechanism for the onset of puberty in mammalian species and its association with metabolic signals remains a challenging enigma.

Acknowledgement

The authors thank Dr. Nancy Levin at Amgen Inc. and Dr. Scott Weigle and Joe Kuijper at ZymoGenetics Inc. for their support. We also thank Dr. Patricia Finn, Matthew Cunningham, John Hohmann, and Shafeena Nurani for their critical evaluation of this manuscript.

References

1 Siegel-Witchel S: CNS lesions, neurological disorders, and puberty in man; in Plant TM, Lee PA (eds): The Neurobiology of Puberty. Bristol, Journal of Endocrinology Ltd, 1995, pp 229–239.
2 Ojeda SR, Urbanski HF: Puberty in the rat; in Knobil E, Neill JD (eds): The Physiology of Reproduction. New York, Raven Press, 1994.
3 Wade GN, Schneider JE: Metabolic fuels and reproduction in female mammals. Neurosci Biobehav Rev 1992;16:235–272.

4 Glass AR, Swerdloff RS: Nutritional influences on sexual maturation in the rat. FASEB J 1980; 39:2360–2364.
5 Wilen R, Naftolin P: Pubertal food intake, body length, weight, and composition in the well fed female rat. Pediatr Res 1977;11:701–703.
6 Holehan AM, Merry BJ: The control of puberty in the dietary restricted female rat. Mech Age Dev 1985;32:179–191.
7 Bronson F: Food-restricted, prepubertal, female rats: Rapid recovery of luteinizing hormone pulsing with excess food, and full recovery of pubertal development with gonadotropin-releasing hormone. Endocrinology 1986;118:2483–2487.
8 Bronson FH: Puberty in female rats: Relative effect of exercise and food restriction. Am J Physiol 1987;252:R140–R144.
9 Bronson FH, Heideman PD: Short-term hormonal responses to food intake in peripubertal female rats. Am J Physiol 1990;259:R25–R31.
10 Warren MP: Effects of undernutrition on reproductive function in the human. Endocr Rev 1983; 4:363–377.
11 Cameron J, Nosbisch C: Suppression of pulsatile luteinizing hormone and testosterone secretion during short term food restriction in the adult male rhesus monkey (*Macaca mulatta*). Endocrinology 1991;128:1532–1540.
12 Kennedy GC, Mitra J: Body weight and food intake as initiating factors for puberty in the rat. J Physiol 1963;166:408–418.
13 Frisch RE, Revelle R: Height and weight at menarche and a hypothesis of critical body weights and adolescent events. Science 1970;169:397–398.
14 Frisch RE, Revelle R, Cook S: Height, weight and age at menarche and the 'critical weight' hypothesis. Science 1971;174:1148–1149.
15 Crawford JD, Osler DC: Body composition at menarche: The Frisch-Revelle hypothesis revisited. Pediatrics 1975;56:449–458.
16 Frisch RE: Pubertal adipose tissue: Is it necessary for normal sexual maturation? Evidence from the rat and human female. FASEB 1980;39:2395–2400.
17 Steiner R, Cameron J, McNeill T, Clifton D, Bremner W: Metabolic signals for the onset of puberty; in Norma R (ed): Neuroendocrine Aspects of Reproduction. New York, Academic Press, 1983, pp 183–227.
18 Cameron JL, Hansen PD, Mcneill DK, Koerker DJ, Clifton DK, Rogers KV, Bremner WJ, Steiner RA: Metabolic cues for the onset of puberty in primate species; in Givens JR, Venturol S (eds): Adolescence in Females. New York, Year Book, 1985, pp 59–78.
19 I'Anson H, Foster DL, Foxcroft GR, Booth PJ: Nutrition and reproduction. Oxford Rev Reprod Biol 1991;13:239–311.
20 Steiner RA: Lords and ladies leapin' on leptin. Endocrinology 1996;137:4533–4535.
21 Zhang Y, Proenca R, Maffei M, Barone M, Leopold L, Friedman JM: Positional cloning of the mouse *obese* gene and its human homologue. Nature 1994;372:425–432.
22 Moinat M, Deng C, Muzzin P, Assimacopoulos-Jeannet F, Seydoux J, Dulloo AG, Giacobino JP: Modulation of *obese* gene expression in rat brown and white adipose tissues. FEBS Lett 1995;373: 131–134.
23 Masuzaki H, Ogawa Y, Sagawa N, Hosoda K, Matsumoto T, Mise H, Nishimura H, Yoshimasa Y, Tanaka I, Mori T, Nakao K: Nonadipose tissue production of leptin: Leptin as a novel placenta-derived hormone in humans. Nat Med 1997;3:1029–1033.
24 Hoggard N, Hunter L, Duncan JS, Williams LM, Trayhurn P, Mercer JG: Leptin and leptin receptor mRNA and protein expression in the murine fetus and placenta. Proc Natl Acad Sci USA 1997; 94:11073–11078.
25 Bado A, Levasseur S, Attoub S, Kermorgant S, Laigneau J, Bortoluzzi M, Moizo L, Lehy T, GuerreMillo M, LeMarchandBrustel Y, Lewin M: The stomach is a source of leptin. Nature 1998; 394:790–793.
26 Maffei M, Hallas J, Ravussin E, Pratlet RE, Lee GH, Zhang Y, Fei H, Kim S, Lallone R, Ranganathan S, Kern PA, Friedman JM: Leptin levels in human and rodent: Measurement of plasma leptin and *ob* RNA in obese and weight-reduced subjects. Nat Med 1995;1:1155–1161.

27 Considine RV, Sinha MK, Heiman ML, Kriauciunas A, Stephens TW, Nyce MR, Ohannesian JP, Marco CC, McKee LJ, Bauer TL, Caro JF: Serum immunoreactive-leptin concentrations in normal-weight and obese humans. N Engl J Med 1996;334:292–295.

28 Frederich RC, Hamann A, Anderson S, Lollmann B, Lowell BB, Flier JS: Leptin levels reflect body lipid content in mice: Evidence for diet-induced resistance to leptin action. Nat Med 1995;1: 1311–1314.

29 Igel M, Kainulainen H, Brauers A, Becker W, Herberg L, Joost HG: Long term and rapid regulation of *ob* mRNA levels in adipose tissue from normal (Sprague-Dawley rats) and obese (*db/db* mice, *fa/fa* rats) rodents. Diabetologia 1996;39:758–765.

30 Boden G, Chen X, Mozzoli M, Ryan I: Effect of fasting on serum leptin in normal human subjects. J Clin Endocrinol Met 1996;81:3419–3423.

31 Kolaczynski JW, Considine RV, Ohannesian J, Marco C, Opentanova I, Nyce MR, Myint M, Caro JF: Response of leptin to short-term fasting and refeeding in humans: A link with ketogenesis but not ketones themselves. Diabetes 1996;45:1511–1515.

32 Kolaczynski JW, Ohannesian JP, Considine RV, Marco CC, Caro JF: Response of leptin to short-term and prolonged overfeeding in humans. J Clin Endocrinol Metab 1996;81:4162–4165.

33 Matson CA, Wiater MF, Weigle DS: Leptin and the regulation of body adiposity: A critical review. Diabetes Rev 1996;4:488–508.

34 Considine RV, Caro JF: Leptin and the regulation of body weight. Int J Biochem Cell Biol 1997; 29:1255–1272.

35 Ingalls AH, Dickie MM, Snell GD: Obese, a new mutation in the house mouse. J Hered 1950;41: 317–318.

36 Zheng D, Wooter MH, Zhou Q, Dohm GL: The effect of exercise on ob gene expression. Biochem Biophys Res Commun 1996;225:747–750.

37 Pelleymounter MA, Cullen MJ, Baker MB, Hecht R, Winters D, Boone T, Collins F: Effects of the *obese* gene product on body weight regulation in *ob/ob* mice. Science 1995;269:540–543.

38 Halaas JL, Gajiwala KS, Margherita M, Cohen SL, Chait BT, Rabinowitz D, Lallone RL, Burley SK, Friedman JM: Weight-reducing effects of the plasma protein encoded by the *obese* gene. Science 1995;269:543–546.

39 Campfield LA, Smith FJ, Guisez Y, Devos R, Burn P: Recombinant mouse OB protein: Evidence for a peripheral signal linking adiposity and central neural networks. Science 1995;269:546–549.

40 Hummel KP, Dickie MM, Coleman DL: Diabetes, a new mutation in the mouse. Science 1966; 153:1127–1128.

41 Zucker LM, Zucker TF: Fatty, a new mutation in the rat. J Hered 1961;52:275–278.

42 Cheung C, Clifton D, Steiner R: Leptin and its action on reproduction; in Bray G (ed): Nutrition and Reproduction. Atlanta, LSU Press, 1998.

43 Swerdloff RS, Batt RA, Bray GA: Reproductive hormonal function in the genetically obese (ob/ob) mouse. Endocrinology 1976;98:1359–1364.

44 Swerdloff RS, Paterson M, Vera A, Batt RAL, Heber D, Bray GA: The hypothalamic-pituitary axis in genetically obese (ob/ob) mice: Response to luteinizing hormone-releasing hormone. Endocrinology 1978;103:542–547.

45 Batt RAL, Everard DM, Gillies G, Wilkinson M, Wilson CA, Yeo TA: Investigation into the hypogonadism of the obese mouse (genotype ob/ob). J Reprod Fert 1982;64:363–371.

46 Barash IA, Cheung CC, Weigle DS, Ren H, Kramer JM, Fallon M, Kabigting EB, Kuijper JL, Clifton DK, Steiner RA: Leptin is a metabolic signal to the reproductive system. Endocrinology 1996;137:3144–3147.

47 Chehab FF, Lim ME, Lu R: Correction of the sterility defect in homozygous obese female mice by treatment with the human recombinant leptin. Nat Genet 1996;12:318–320.

48 Chehab FF, Mounzih K, Lu R, Lim ME: Early onset of reproductive function in normal female mice treated with leptin. Science 1997;275:88–90.

49 Ahima RS, Prabakaran D, Mantzoros C, Qu D, Lowell B, Maratos-Flier E, Flier JS: Role of leptin in the neuroendocrine response to fasting. Nature 1996;382:250–252.

50 Finn PD, Cunningham MJ, Pau K-YF, Spies HG, Clifton DK, Steiner RA: The stimulatory effect of leptin on the neuroendocrine axis of the monkey. Endocrinology 1998;139:4652–4662.

51 Cheung CC, Thornton JE, Kuijper KL, Weigle DS, Clifton DK, Steiner RA: Leptin is a metabolic gate for the onset of puberty in the female rat. Endocrinology 1997;138:855–858.

52 Gruaz NM, Lalaoui M, Pierroz DD, Sizonenko PC, Blum WF, Aubert ML: Chronic administration of leptin into the lateral ventricle induces sexual maturation in severely food-restricted female rats. J Neuroendocrinol 1998;10:627–633.

53 Banks WA, Kastin AJ, Huang W, Jaspan JB, Maness LM: Leptin enters the brain by a saturable system independent of insulin. Peptides 1996;17:305–311.

54 Schwartz MW, Peskind E, Raskind M, Boyko EJ, Porter D Jr: Cerebrospinal fluid leptin levels: Relationship to plasma levels and to adiposity in humans. Nat Med 1996;2:589–593.

55 Mercer JG, Hoggard N, Williams LM, Lawrence CB, Hannah LT, Trayhurn P: Localization of leptin receptor mRNA and the long form splice variant (Ob-Rb) in mouse hypothalamus and adjacent brain regions by in situ hybridization. FEBS Lett 1996;387:113–116.

56 Guan XM, Hess JF, Yu H, Hey PJ, van der Ploeg LHT: Differential expression of mRNA for leptin receptor isoforms in the rat brain. Mol Cell Endocrinol 1997;133:1–7.

57 Elmquist J, Bjorbaek C, Ahima R, Flier J, Saper C: Distribution of leptin receptor mRNA isoforms in the rat brain. J Comp Neurol 1998;395:535–547.

58 Vaisse C, Halaas JL, Horvath CM, Darnell JE Jr, Stoffel M, Friedman JM: Leptin activation of Stat3 in the hypothalamus of wild-type and *ob/ob* mice but not *db/db* mice. Nat Genet 1996;14: 95–97.

59 Nagatani S, Guthikonda P, Thompson RC, Tsukamura H, Maeda K, Foster DL: Evidence for GnRH regulation by leptin: Leptin administration prevents reduced pulsatile LH secretion during fasting. Neuroendocrinology 1998;67:370–376.

60 Yu WH, Kimura M, Walczewska A, Karanth S, McCann SM: Role of leptin in hypothalamic-pituitary function. Proc Natl Acad Sci USA 1997;94:1023–1028.

61 Ahima RS, Dushay J, Flier S, Prabakaran D, Flier JS: Leptin accelerates the onset of puberty in normal female mice. J Clin Invest 1997;99:391–395.

62 Devaskar SU, Ollesach C, Rajakumar RA, Rajakumar PA: Developmental changes in *ob* gene expression and circulating leptin peptide concentrations. Biochem Biophys Res Commun 1997;238: 44–47.

63 Rayner DV, Dalgliesh GD, Duncan JS, Hardie LJ, Hoggard N, Trayhurn P: Postnatal development of the *ob* gene system: Elevated leptin levels in suckling *fa/fa* rats. Am J Physiol 1997;273:R446–R450.

64 Ahima RS, Prabakaran D, Flier JS: Postnatal leptin surge and regulation of circadian rhythm of leptin by feeding: Implications for energy homoeostatis and neuroendocrine function. J Clin Invest 1998;101:1020–1027.

65 Plant TM, Durrant AR: Circulating leptin does not appear to provide a signal for triggering the initiation of puberty in the male rhesus monkey (*Macaca Mulatta*). Endocrinology 1997;138: 4505–4508.

66 Urbanski HF, Pau KYF: A biphasic developmental pattern of circulating leptin in the male rhesus macaque (*Macaca mulatta*). Endocrinology 1998;139:2284–2286.

67 Hassink SG, Sheslow DV, de Lancey E, Opentanova I, Considine RV, Caro JF: Serum leptin in children with obesity: Relationship to gender and development. Pediatrics 1996;98:201–203.

68 Garcia-Mayor RV, Andrade MA, Rios M, Lage M, Dieguez C, Casanueva FE: Serum leptin levels in normal children: Relationship to age, gender, body mass index, pituitary-gonadal hormones, and pubertal stage. J Clin Endocrinol Metab 1997;82:2849–2855.

69 Blum WF, Englaro P, Hanitsch S, Juul A, Hertel NT, Muller J, Skakkebaek NE, Heiman ML, Birkett M, Attanasio AM, Kiess W, Rascher W: Plasma leptin levels in healthy children and adolescents: Dependence on body mass index, body fat mass, gender, pubertal stage, and tetosterone. J Clin Endocrinol Metab 1997;82:2904–2910.

70 Wabitsch M, Blum W, Muche R, Braun M, Hube F, Rascher W, Heinze E, Teller W, Hauner H: Contribution of androgens to the gender difference in leptin production in obese children and adolescents. J Clin Invest 1997;100:808–813.

71 Lahlou N, Landais P, De Boissieu D, Bougneres PF: Circulating leptin in normal children and during the dynamic phase of juvenile obesity: Relation to body fatness, energy metabolism, caloric intake, and sexual dimorphism. Diabetes 1997;46:989–993.

72 Clayton PE, Gill MS, Hall CM, Tillmann V, Whatmore AJ, Proce DA: Serum leptin through childhood and adolescence. Clin Endocrinol 1997;46:727–733.

73 Carlsson B, Ankarberg C, Rosberg S, Norjavaara E, Albertsson-Wikland K, Carlsson LMS: Serum leptin concentrations in relation to pubertal development. Arch Dis Child 1997;77:396–400.

74 Arslanian S, Suprasongsin C, Kalhan SC, Drash AL, Brna R, Janosky JE: Plasma leptin in children: Relationship to puberty, gender, body composition, insulin sensitivity, and energy expenditure. Metabolism 1998;47:309–312.

75 Pombo M, Herrera-Justiniano E, Considine RV, Hermida RC, Galvex MJ, Martin T, Barreiro J, Casanueva FF, Dieguez C: Nocturnal rise of leptin in normal prepubertal and pubertal children and in patients with perinatal stalk-transection syndrome. J Clin Endocrinol Metab 1997;82:2751–2754.

76 Argente J, Barrios V, Chowen JA, Sinha MK, Considine RV: Leptin plasma levels in healthy Spanish children and adolescents, children with obesity, and adolescents with anorexia nervosa and bulimia nervosa. J Pediatr 1997;131:833–838.

77 Matkovic V, Ilich JZ, Skugor M, Badenhop NE, Goel P, Clairmont A, Klisovic D, Nahhas RW, Landoll JD: Leptin is inversely related to age at menarche in human females. J Clin Endocrinol Metab 1997;82:3239–3245.

78 Yoneda N, Saito S, Kimura M, Yamada M, Iida M, Irahara M, Shima K, Aono T: The influence of ovariectomy on ob gene expression in rats. Horm Metab Res 1998;30:263–265.

79 Shimizu H, Shimomura Y, Nakanishi Y, Futawatari T, Ohtani K, Sato N, Mori M: Estrogen increases in vivo leptin production in rats and human subjects. J Endocrinol 1997;154:285–292.

80 Mantzoros CS, Flier JS, Rogol AD: A longitudinal assessment of hormonal and physical alterations during normal puberty in boys. V. Rising leptin levels may signal the onset of puberty. J Clin Endocrinol Metab 1997;82:1066–1070.

81 Jockenhovel F, Blum WF, Vogel E, Englaro P, Muller-Weland D, Rainwein D, Rascher W, Krone W: Testosterone substitution normalizes elevated serum leptin levels in hypogonadal men. J Clin Endocrinol Metab 1997;82:2510–2513.

82 Palmert MR, Radovick S, Boepple PA: The impact of reversible gonadal sex steroid suppression on serum leptin concentrations in children with central precocious puberty. J Clin Endocrinol Metab 1998;83:1091–1096.

83 Strobel A, Issad T, Camoin L, Ozata M, Strosberg AD: A leptin missense mutation associated with hypogonadism and morbid obesity. Nat Genet 1998;18:213–215.

84 Montague CT, Farooqi IS, Whitehead JP, Soos MA, Rau H, Wareham NJ, Sewter CP, Digby JE, Mohammed SN, Hurst JA, Cheetham CH, Earley AR, Barnett AH, Prins JB, O'Rahilly S: Congenital leptin deficiency is associated with severe early-onset obesity in humans. Nature 1997; 387:903–908.

85 Clement K, Vaisse C, Lahlou N, Cabroll S, Pelloux V, Cassuto D, Gourmelen M, Dina C, Chambaz J, Lacrote J, Basdevant A, Bougeneres P, Lebouc Y, Frogue P, Guy-Grand B: A mutation in the human leptin receptor gene causes obesity and pituitary dysfunction. Nature 1998;392:398–401.

86 Cheung CC, Clifton DK, Steiner RA: Proopiomelanocortin neurons are direct targets for leptin in the hypothalamus. Endocrinology 1997;138:4489–4492.

87 Mercer JG, Hoggard N, Williams LM, Lawrence CB, Hannah LT, Morgan PJ, Trayhurn P: Coexpression of leptin receptor and preproneuropeptide Y mRNA in arcuate nucleus of mouse hypothalamus. J Neuroendocrinol 1996;8:733–735.

88 Hakansson ML, Brown H, Ghilardi N, Skoda RC, Meister B: Leptin receptor immunoreactivity in chemically defined target neurons in the hypothalamus. J Neurosci 1998;18:559–572.

89 Kalra SP, Sahu A, Kalra PS, Crowley WR: Hypothalamic neuropeptide Y: A circuit in the regulation of gonadotropin secretion and feeding behavior. Ann NY Acad Sci 1990;611:273–283.

90 Gruaz NM, Pierroz DD, Rohner-Jeanrenaud F, Sizonenko PC, Aubert ML: Evidence that neuropeptide Y could represent a neuroendocrine inhibitor of sexual maturation in unfavorable metabolic conditions in the rat. Endocrinology 1993;133:1891–1894.

91 Pierroz DD, Gruaz NM, d'Alleves V, Aubert ML: Chronic administration of neuropeptide Y into the lateral ventricles starting at 30 days of life delays sexual maturation in the female rat. Neuroendocrinology 1995;61:293–300.

92 Stephens TW, Basinski M, Bristow PK, Bue-Valleskey JM, Burgett SG, Craft L, Hale J, Hoffmann J, MacKellar W, Rosteck Jr PR, Schoner B, Smith D, Tinsley FC, Zhang X, Helman M: The role of neuropeptide Y in the antiobesity action of the obese gene product. Nature 1995;12:530–532.
93 Schwartz MW, Seeley RJ, Campfield LA, Burn P, Baskin DG: Identification of targets of leptin action in rat hypothalamus. J Clin Invest 1996;98:1101–1106.
94 Erickson JC, Clegg KE, Palmiter RD: Sensitivity to leptin and susceptibility to seizures of mice lacking neuropeptide Y. Nature 1996;381:415–418.
95 Thornton JE, Cheung CC, Clifton DK, Steiner RA: Regulation of hypothalamic proopiomelanocortin mRNA by leptin in *ob/ob* mice. Endocrinology 1997;138:5063–5066.
96 Scimonelli T, Celis ME: A central action of alpha-melanocyte-stimulating hormone on serum levels of LH and prolactin in rats. J Endocrinol 1990;124:127–132.
97 Caballero C, Celis ME: The effect of the blockade of alpha-melanocyte-stimulating hormone on LH release in the rat. J Endocrinol 1993;137:197–202.

Robert A. Steiner, PhD, Department of Physiology and Biophysics,
University of Washington, Box 357290, Seattle, WA 98195-7290 (USA)
Tel. +1 206 543 3915, Fax +1 206 685 0619, E-Mail steiner@u.washington.edu

Ur E (ed): Neuroendocrinology of Leptin.
Front Horm Res. Basel, Karger, 2000, vol 26, pp 106–125

The Brain Is a Source of Leptin

Michael Wilkinson[a, b], *Barbara Morash*[c], *Ehud Ur* [a, c]

Departments of [a] Obstetrics and Gynaecology, [b] Physiology and Biophysics and [c] Division of Endocrinology, Faculty of Medicine, Dalhousie University, Halifax, Nova Scotia, Canada

Introduction

Since its discovery in 1994, studies on the hormone leptin have generated over 1,400 MEDLINE citations and several reviews of this expanding literature are already available [Sawchenko, 1998; Elmquist et al., 1998; Dagogo-Jack, 1999]. Other contributors to the present volume provide detailed updates of the neuroendocrinological role of leptin in the regulation of appetite and body weight (see chapters 1 and 2). The hypothalamus is evidently the seat of a complex leptin signaling system that participates in the control of appetite and body weight through interactions with melanocortins, CART, CRF, galanin, neurotensin, NPY, hypocretins/orexins and steroid hormones [Elmquist et al., 1999; Kalra et al., 1999].

Leptin receptors exist as several splice variants [Bjørbaek et al., 1997; Uotani et al., 1999] and much is known about their hypothalamic distribution in rodent, human and monkey brain [chapter 3; and references therein; Guan et al., 1997; Shioda et al., 1998; Finn et al., 1998]. There is also an extensive distribution of these receptors in many extrahypothalamic brain regions, including cerebral cortex, cerebellum, substantia nigra, medial and lateral geniculate nuclei, hippocampus, olfactory bulb, dorsal raphé and inferior olive nuclei, nucleus of the solitary tract and the dorsal motor nucleus of the vagus nerve. In addition, leptin receptor isoforms are also found in rat brain microvessels [Bjørbaek et al., 1998].

Leptin is a large peptide (16 kD) that cannot readily cross the blood-brain barrier. This raises the paradox of why OBR are so widely localized in the brain where they are likely to be inaccessible to circulating leptin. Arguments for a specific transport system have been made (Elmquist, 1998; and see below for discussion).

We have hypothesized that central leptin receptors – with the exception of those in the basal hypothalamus – are accessible only to brain-derived leptin.

In the following discussion we will explore the possibility that leptin is an endogenous brain ligand for central leptin receptors. This discussion will take into account prevailing theories and also provide evidence for leptin gene expression in the CNS.

What Is the Role of Nonhypothalamic Leptin Receptors?

The heterogeneous distribution of CNS receptors in regions such as cerebellum, cerebral cortex and hippocampus suggests that leptin might modulate neural systems which are distinct from those involved in body weight regulation. This issue is complicated by a relative absence of information on the cellular localization of leptin receptors (OBR), i.e. are OBR found in neurons or in glia, or both? In the hypothalamus for example some, but not all, OBR immunopositive cells contain neuronal markers such as CRF, oxytocin, NPY and estrogen receptors [Håkansson et al., 1998; Baskin et al., 1999; Diano et al., 1998a]. Ultrastructural studies have also revealed OBR labelling of both neurons and glia [de Matteis and Cinti, 1998; Diano et al., 1998b]. Double-labelling studies with neuron and glial-specific markers are necessary to definitively establish the identity of OBR-positive cells. Non-neuronal cells elsewhere are known to possess OBR mRNA, e.g. adipose tissue [Kieler et al., 1998], spleen, testes, kidney and lung [Hoggard et al., 1997].

The preponderance of the evidence would suggest that leptin is unlikely to cross the bloodbrain barrier to reach nonhypothalamic OBR. For example, attempts to localize peripherally injected [^{125}I]-leptin in brain have failed to show central labelling except at the choroid plexus and the arcuate-median eminence region of the hypothalamus [Van Heek et al., 1996; Malik and Young 1996; Banks et al., 1996]. Even label injected intracerebroventricularly is quickly cleared into the blood and fails to label even the arcuate nucleus [Maness et al., 1998]. These data indicate that OBR located at sites within the brain, distant from the ventricles, such as those in hippocampus, substantia nigra and cerebellum, are unlikely to be accessible to circulating leptin. This is consistent with neuroanatomical studies showing that entry of large molecules into brain parenchyma occurs only at the circumventricular regions such as median eminence [Broadwell and Sofroniew, 1993; Bouchard and Bosler, 1986]. Even at these sites peptides may well breach the BBB, but penetrate only a short distance into the brain parencyhma (e.g. from median eminence to arcuate nucleus). A similar conclusion was reached by Herkenham et al.

[1998] in a detailed study of *c-fos* expression following peripheral injection of IL-1 [see also Elmquist et al., 1997]. In addition, Sinha et al. [1996] have suggested that most circulating leptin is in the form of a leptin-binding protein complex which is even less likely to enter the brain. Indeed in rhesus monkeys, leptin reduces food intake only when administered centrally, suggesting that increases in circulating leptin within the physiological range may not be sufficient to acutely regulate food intake [Ramsey et al., 1998; Tang-Christensen et al., 1999]. A counter argument is that a specific transport system for leptin may be located in microvessels and choroid plexus [Golden et al., 1997; Bjørbaek et al., 1998].

Is There a Transport System for Leptin Entry Into the Brain?

The presence of a specific transport system, operating through the short leptin receptor isoform, remains unproven. However, even if transport of leptin by short form receptors does exist, it does not explain why peripheral injection of high specific activity [^{125}I] leptin fails to label extrahypothalamic areas other than choroid plexus [Banks et al., 1996]. It is conceivable that the labelling of leptin with [^{125}I] might compromise binding and transport of the leptin molecule. However this seems unlikely in view of the binding experiments reported by Uotani et al. [1999]. These authors successfully used radioiodinated leptin to label and quantify both long- and short-form (OBRb and OBRa, respectively) receptors stably expressed in Chinese hamster ovary cells. These receptors had K_d values in the nanomolar range and were also capable of internalizing [^{125}I]-leptin via coated pits, confirming that in this system the labelled leptin species was able to bind normally to its cognate receptor.

These experiments [Uotani et al., 1999] are notable for an additional reason. In this study, the OBRa were more efficient in mediating rapid internalization and degradation of the [^{125}I] leptin than were the long form (OBRb) receptors. The data suggest that OBRa located in brain microvessels may not act to transport intact leptin into the brain at all but rather ensure that the hormone is degraded before this occurs. Such a mechanism – an 'enzymatic' or 'metabolic' blood-brain barrier – has previously been postulated for neuroactive peptides [Meisenberg and Simmons, 1983; Goldstein and Betz, 1986]. It is also possible that the presence of OBR in microvessels is related to leptin's action as an angiogenic factor [Bouloumie et al., 1998; Sierra-Honigmann et al., 1998].

Leptin Elevates FOS and STAT 3

That certain populations of central OBR are inaccessible to peripheral injections of leptin can also be inferred from experiments using *c-fos* expression. FOS protein-like immunoreactivity (FLI) has been used extensively to localize *c-fos* expression as a marker of cellular activation [Hughes and Dragunow, 1995; Natarajan and Wilkinson, 1997]. We predicted that peripheral injection of leptin should increase FLI in basal hypothalamus but not, for example, in hippocampus and cerebral cortex. This is indeed the case, since we observed leptin-induced *c-fos* expression only in basal hypothalamus (fig. 1) and these data are largely in agreement with other reports [Woods and Stock 1996; Seeley et al., 1997; Van Dijk et al., 1996; Wang et al., 1998; Elmquist et al., 1997]. Interestingly, Elmquist et al. [1997] and Kalra et al. [1999] did not observe leptin-induced FLI in the arcuate nucleus.

A *lack* of *c-fos* expression does not necessarily imply an absence of leptin-induced cell activation. Equally, *c-fos* may not be expressed in all cells that have leptin receptors, i.e. leptin injection could act along a multisynaptic pathway involving several neurotransmitters. A critical experiment would be to determine whether *c-fos* expression is increased following microinjection of leptin into hippocampus and cerebellum. An alternative approach is to determine the effect of peripheral leptin on OBR_b (the long form) which are coupled to the *janus* kinase (JAK)-STAT pathway. Leptin target neurons in brain contain STAT 3-ir (immunoreactivity) [Håkkansson and Meister, 1998] and leptin injection activates STAT 3 in hypothalamus [Vaisse et al., 1996]. Our experiments show that, as with *c-fos*, leptin increases STAT 3-ir in hypothalamus but not in hippocampus, cerebral cortex or substantia nigra (fig. 2). This again implies that leptin does not readily penetrate the blood brain barrier.

In summary, the available evidence supports our contention that peripherally derived leptin is unlikely to be the ligand for those leptin receptors located in extrahypothalamic brain regions. This raises the possibility that leptin, or a leptin-like molecule, of CNS origin, is the ligand for central OBR. An analogous case has been made for a putative role for insulin in brain [Devaskar et al., 1994; Schechter et al., 1998]. However, in the case of leptin the hypothesis is of greater significance since the brain is the principal target for leptin of peripheral origin.

Leptin Gene Expression in Rat Brain

To determine whether leptin mRNA was expressed in the rat brain, we used reverse transcriptase polymerase chain reaction (RT-PCR) with intron

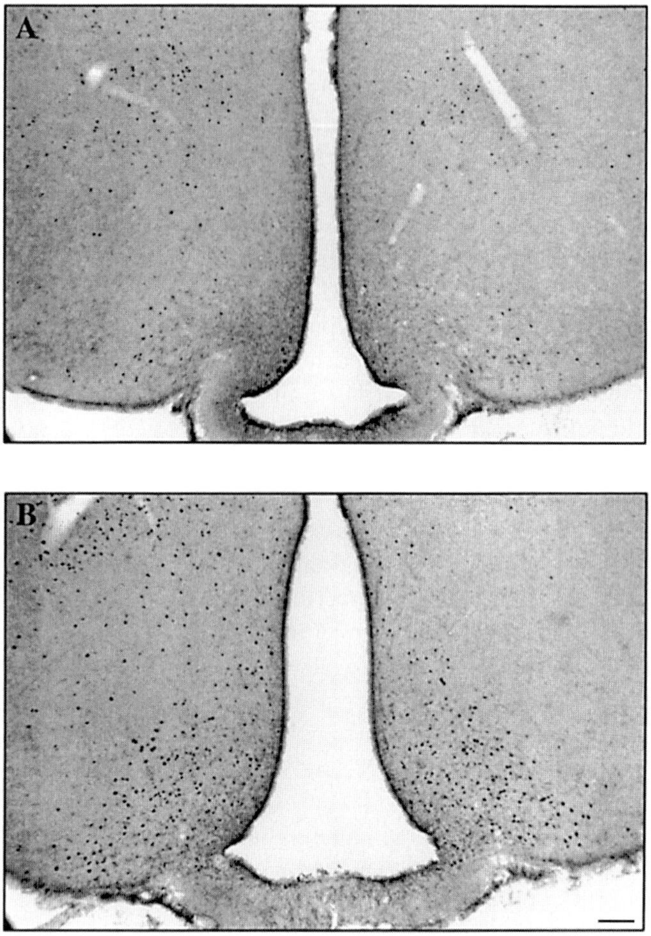

Fig. 1. *C-fos* response to leptin in rat hypothalamus. Fos-like immunoreactivity (FLI) in basal hypothalamus of adult female rats (126–150 g) given saline (*A*) or leptin (*B*; 1 mg/kg s.c.). Note that leptin-induced FLI is localized largely to the arcuate nucleus with no staining in the median eminence. Rats were perfused with paraformaldehyde (4%) 2 h postinjection. These photomicrographs are typical of numerous experiments. Scale bar = 100 μm.

spanning primers directed at a 217-bp region spanning exons 2 and 3 of the leptin transcript. A product of the expected size (217 bp) was detected in the following rat tissues: fat, cerebral cortex, cerebellum, hypothalamus, and anterior pituitary (fig. 3) indicating that these tissues express leptin mRNA. Leptin mRNA was undetectable in liver and at very low levels in hippocampus

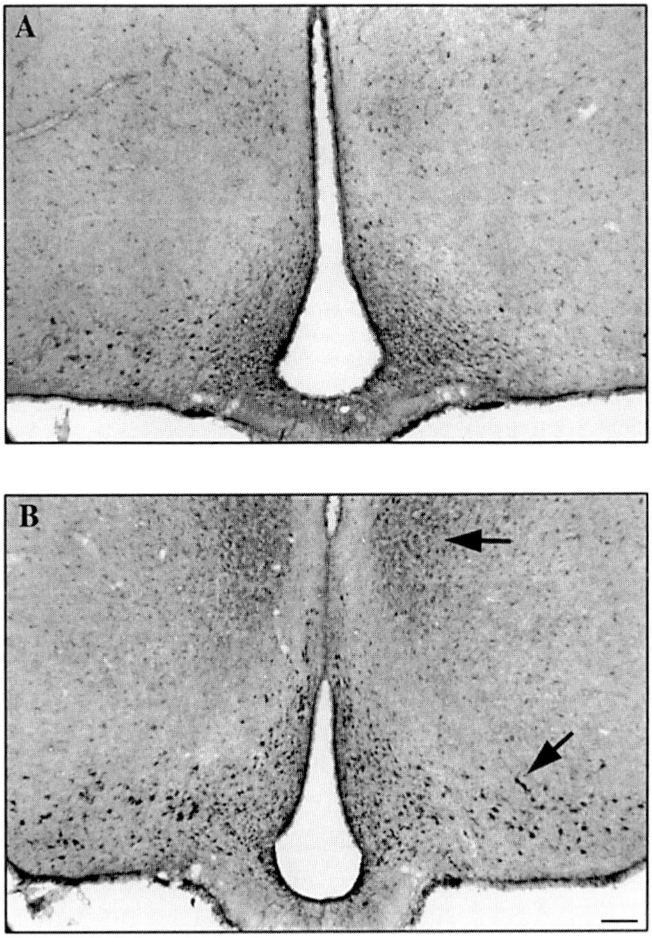

Fig. 2. STAT 3 response to leptin in rat hypothalamus. STAT 3-ir in coronal sections of basal hypothalamus of adult female rat detected with a polyclonal antibody (Santa Cruz; 1:20,000) in response to saline (*A*) or leptin (*B*; 1 mg/kg s.c.). Note the appearance of stained cells in the dorsomedial, periventricular and lateral hypothalamus (arrows). Rats were perfused 2 h postinjection. Scale bar = 100 μm.

under the same conditions. Sequence analysis of the 217-bp amplicon from fat and hypothalamus confirmed 100% homology with the corresponding region of the published sequence for leptin cDNA.

These findings are in contrast with those previously reported in the literature [Zhang et al., 1994]. Possible explanations for this include use of more

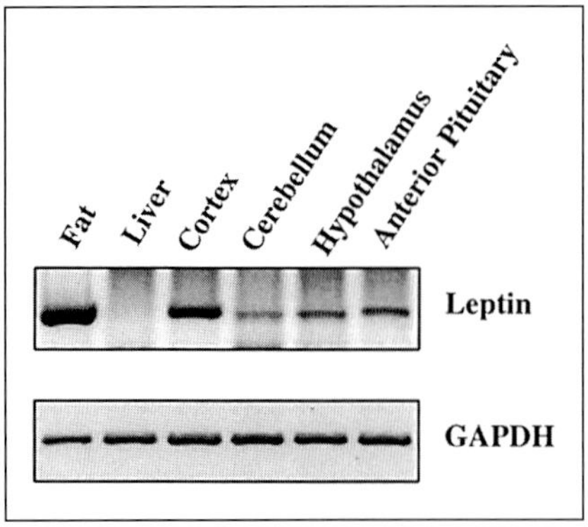

Fig. 3. Detection of leptin mRNA in rat neuronal and peripheral tissues by RT-PCR. Total RNA was reverse transcribed and PCR-amplified (40 cycles). A product of the expected size (217 bp) was detected in various rat tissues including: fat (lane 1), cerebral cortex (lane 3), cerebellum (lane 4), hypothalamus (lane 5), and anterior pituitary (lane 6). No PCR product was detected in liver (lane 2).

input RNA for cDNA synthesis, higher cycle numbers and different primers combined with robust optimization of the PCR reaction conditions. Another important difference between our work and that of Zhang et al. [1994] is their use of mouse tissues. In our hands only a weak signal for the 217-bp product was obtained from mouse fat (40 cycles) and brain tissues (hypothalamus and cerebral cortex) were negative.

It may be argued that detection of leptin gene expression by RT-PCR is of doubtful physiological relevance, and merely reflects contamination of brain tissue with adipocytes. We think this is unlikely in view of our inability to detect expression in liver. In addition, we have observed strong signals for ob mRNA expression in two cell types which are certainly free of adipocytes, viz. C6 glioblastoma and pituitary RC-4B/C cells grown in culture (unpublished observations).

Localization of Leptin Immunoreactivity in Rat and Mouse Brain

There is a large literature devoted to the localization of leptin *receptors* by immunocytochemistry, but little is published on the leptin protein. This is limited to studies by Hoggard et al. [1997] and Wang et al. [1998] who have identified leptin-like immunoreactivity in placenta and skeletal muscle, respectively. The detection of ob mRNA expression in rat brain led us to examine the distribution of leptin-immunoreactivity (leptin-ir) in rat brain sections. We used two polyclonal antibodies raised against peptide sequences at the N-terminus (amino acids 16–34) and the C-terminus (amino acids 137–156) of human leptin (Santa Cruz Biotech: #Y-20 and A-20, respectively). Both antibodies react with leptin of rat, mouse and human origin by Western blotting.

These experiments revealed marked antibody-dependent differences in the localization of leptin-ir in rat and mouse brain sections. The A-20 ab (C-terminus) revealed staining in the rat cerebellum (fig. 4a), light staining in cerebral cortex (fig. 4b) and in arcuate nucleus/periventricular area (fig. 4c) (fig. 4). Immunoreactivity was abolished by omitting the primary antibody or by preadsorbing the primary with the immunizing peptide (Santa Cruz Bio Tech #SC-842 P). In contrast, leptin-ir was completely absent in mouse brain sections (C57 BL) except for distinct tanycyte-like staining in the median eminence (fig. 4d). In rat brain sections leptin-ir is clearly localized to cell nuclei in the ARC/periventricular area but in cortex and cerebellum cytoplasm and cell processes are also labelled.

The Y-20 (N-terminus) antibody provided a markedly different staining pattern. In rat brain, nuclear leptin-ir was widespread, including ARC, ependymal cells lining the third ventricle (fig. 5A), cortex (retrosplenial; piriform) and hippocampus (dentate gyrus) (fig. 5B). In cerebellum the Purkinje cells were not stained but cells were labelled throughout the molecular layer, again unlike the results with the A20 antibody (fig. 5C). Hippocampal cells were also labelled in hamster brain (fig. 5D). Staining was abolished by omitting the primary antibody or by preadsorbing with the Y20 peptide. In mouse brain sections the Y20 antibody revealed impressive leptin-like-ir in ARC, cortex and dentate gyrus plus some striking multipolar cells (e.g. in suprachiasmatic nucleus; fig. 6A) and many densely labelled tanycytes (fig. 6B). Similar Y20 staining was seen in hamster brain sections, including dense labelling in dentate gyrus (not shown).

Recently, Li et al. [1999] reported the presence of leptin-ir in the CNS of normal and diabetic rats. In contrast to our method they used a single antibody raised against a different epitope (N-terminus; amino acids, 25–44) and their rats were treated with intraventricular colchicine 48 h prior to perfusion. Nevertheless, cytoplasmic leptin-ir was detectable in hypothalamus and brainstem. In addition they were able to demonstrate that hypothalamic leptin-ir was insulin-dependent in streptozotocin-induced diabetic rats. These data support our evidence that the leptin gene is expressed in brain tissue.

In summary, using two different antibodies we observed leptin-like immunoreactivity in rat, mouse and hamster brain. An important question relates to the origin of the leptin detected in these studies. For reasons outlined above, we believe that leptin is produced in the brain and, with the exception of the hippocampus, leptin-ir is observed in those brain regions where ob mRNA is detectable by RT-PCR. However, the heavy staining of ependymal and tanycyte-like cells also suggests that uptake of leptin from CSF is possible. In these cells the immunoreactivity is clearly cytoplasmic whereas elsewhere (e.g. ARC

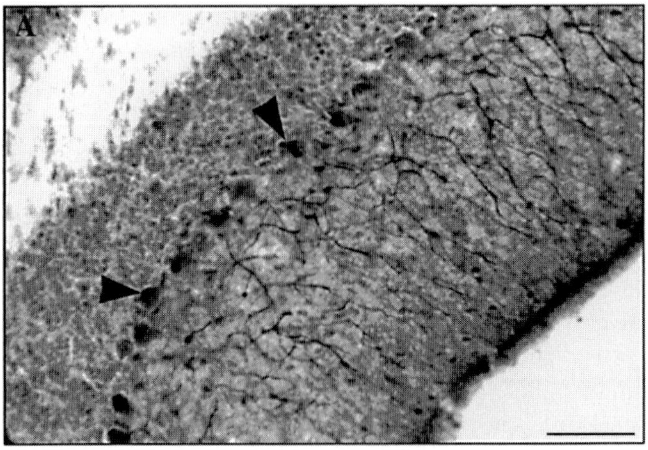

4A

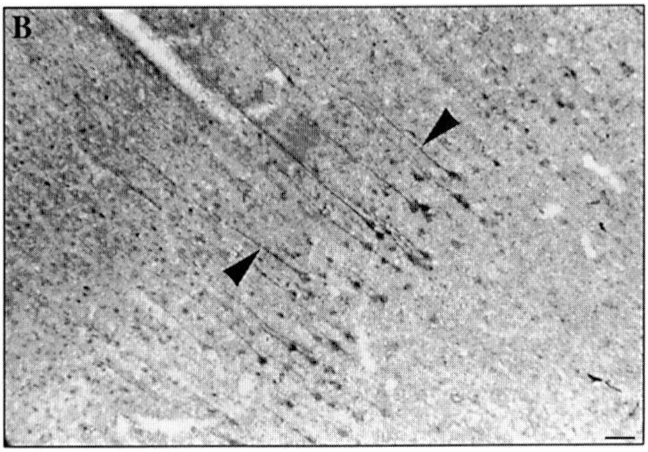

4B

Fig. 4. Immunohistochemical detection of leptin in rat and mouse brain. Coronal sections from fixed brain of adult female rats or mice stained for leptin immunoreactivity using a polyclonal antibody raised against a peptide sequence mapping at the C-terminus of human leptin (Santa Cruz; A20; 1:5,000). *A* Staining in Purkinje cells of the cerebellum. Some fainter labelling was also noted in the cerebral cortex (*B*). *C* Labelled cells in the arcuate nucleus and median eminence. *D* Marked labelling in the mouse median eminence (tanycytes?). Staining was abolished by removing the primary antibody from the staining reaction or by preadsorbing the primary antibody with the peptide antigen. Scale bars = 100 μm.

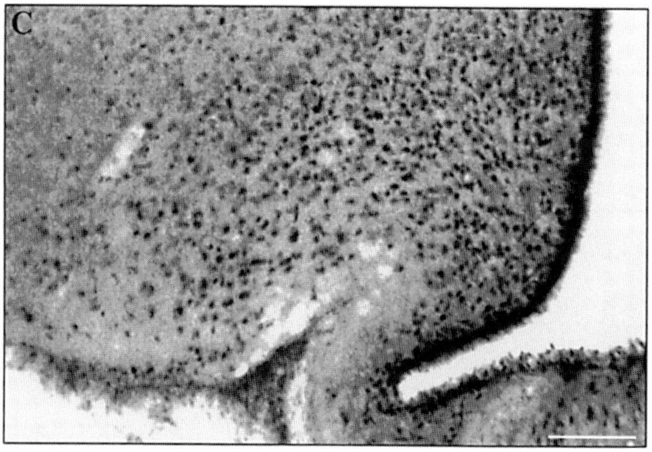

4C

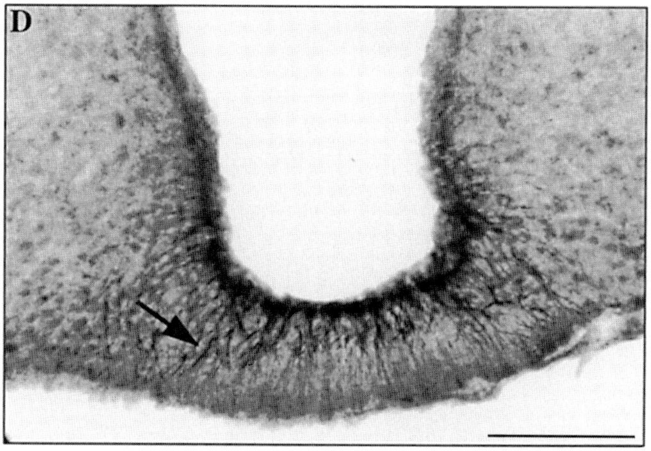

4D

and cortex) it is nuclear, indicating the existence of a *nuclear* leptin receptor. At present it is unclear why two antibodies, raised against different epitopes on the same molecule (leptin), yield distinct labelling patterns. One possibility is that the leptin-ir represents a truncated leptin fragment which contains the N-terminal region of the leptin peptide. An important goal now is to determine the cellular localization of leptin-ir. The type of cell which produces this molecule (glia; neurons) should provide important information on its role in the CNS. For example, could it behave as a neurotransmitter/modulator following release from neuron terminals? Or is it some type of glial signalling factor? (see next section).

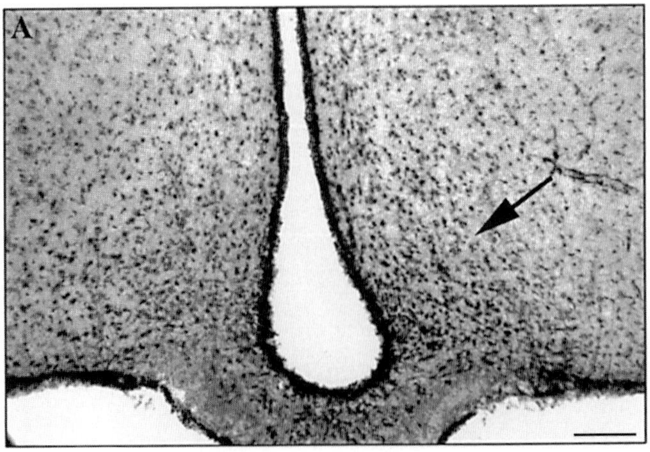

5A

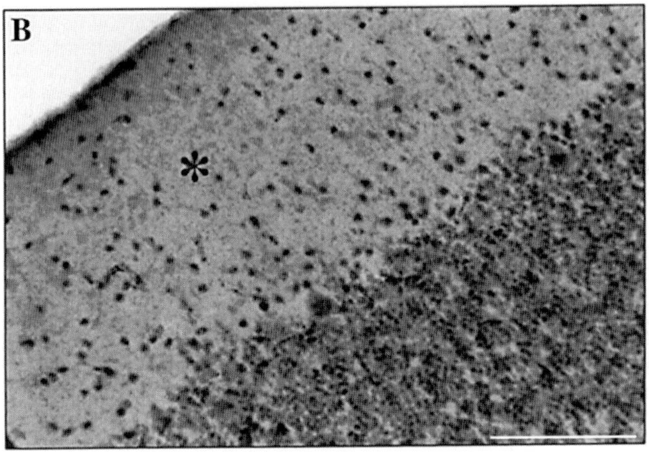

5B

Fig. 5. Immunohistochemical detection of leptin in rat and hamster brain. Coronal sections from fixed brain stained for leptin immunoreactivity using an antibody raised against a peptide from the A-terminus of human leptin (Santa Cruz; #Y-20; 1:5,000). *A* Labelled cells in the arcuate nucleus and median eminence and heavily stained ependymal cells. *B* Staining in the molecular (*) and granular layers of the cerebellum. *C* Marked labelling in dentate gyrus. *D* Hamster hippocampus. The sections treated with antibody preadsorbed with immunizing peptide were not stained (not shown). Scale bars = 100 μm.

Neurobiology of Brain-Derived Leptin

The first report on leptin, in 1994, heralded intensive work on the hormonal signalling pathway that links adipose tissue and the brain. Until recently, it was accepted that adipose tissue represented the only source of leptin.

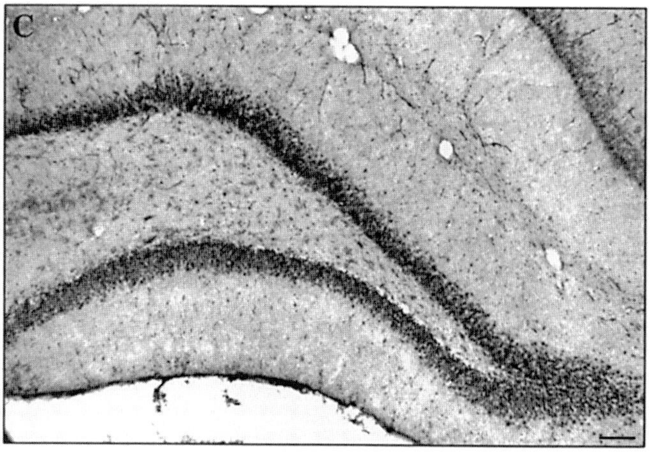

5C

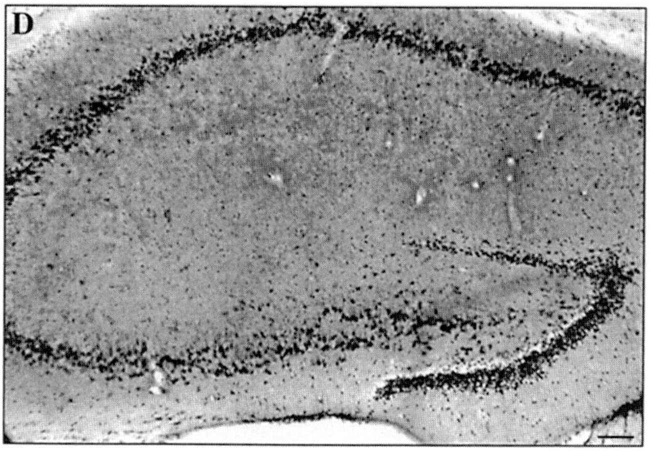

5D

However there is now good evidence for leptin gene expression in placenta, gastric epithelium and skeletal muscle [Hoggard et al., 1997; Bado et al., 1998; Wang et al., 1998] as well as adrenal gland and breast tissue [Bottner et al., 1999; O'Brien et al., 1999]. The brain and pituitary gland should also be added to this list [Morash et al., 1999; Ur et al., 1999].

Our data raise many questions, the most obvious of which is: What is the role of leptin produced locally within the brain? As outlined already above, leptin receptors in the arcuate nucleus are strategically located targets for circulating leptin whereas those inside the BBB are not. The receptors in cerebral cortex and cerebellum should be accessible only to leptin produced in these areas. However, as we have also shown, the hypothalamus expresses

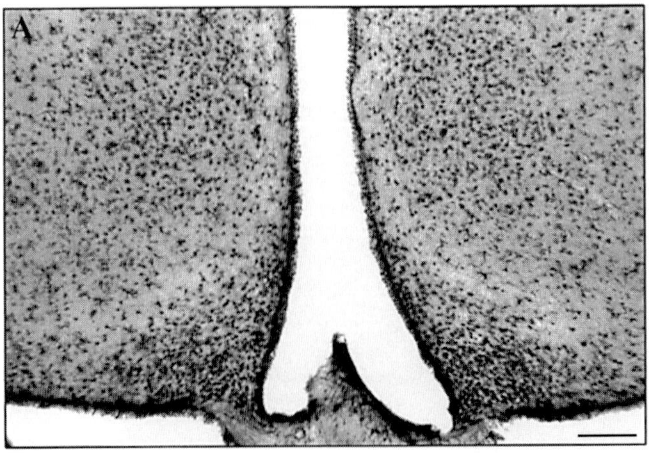

6A

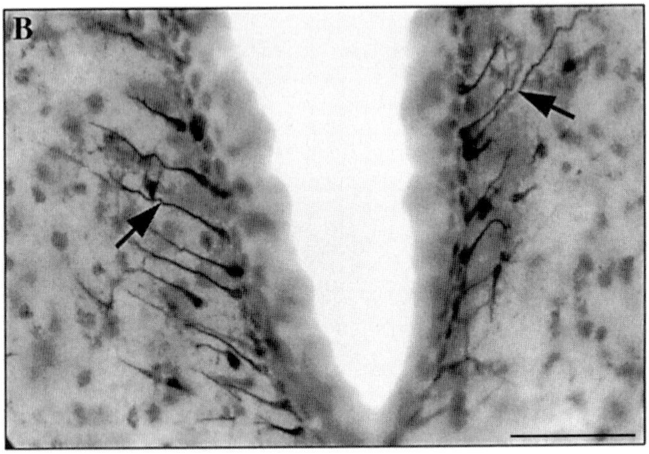

6B

Fig. 6. Leptin-like immunoreactivity in mouse brain. Leptin-like immunoreactivity in C57 black mouse brain using the Y-20 antibody. *A* Leptin-ir arcuate nucleus and basal hypothalamus. *B* Tanycyte staining in the dorsal third ventricular region. *C* Staining of multipolar cells in the suprachiasmatic area. *D* Dense staining in the dentate gyrus and CA1 region of the hippocampus. Omission of the primary antibodies or preadsorption of the antisera with the immunizing peptide completely abolished staining. Scale bars = 100 μm (*A, C, D*) and 500 μm (*B*).

ob mRNA, suggesting that hypothalamic receptors could be responsive to both central and peripheral leptin. Such an occurrence is not without precedent since, for example, ARC and median eminence μ-opioid receptors are exposed to circulating β-endorphin as well as to β-endorphin released from POMC neurons [Mansour et al. 1995].

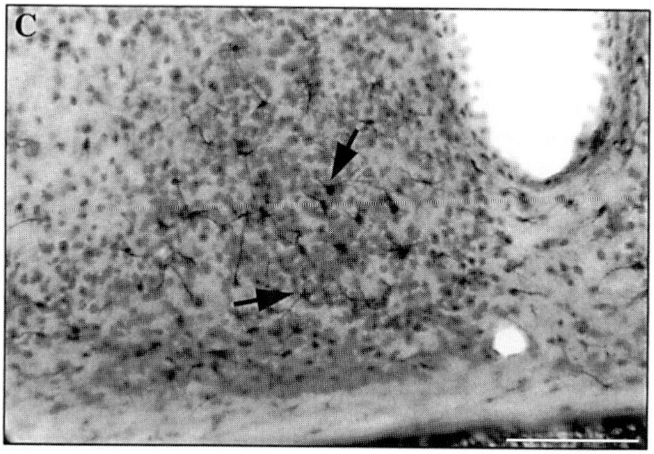

6C

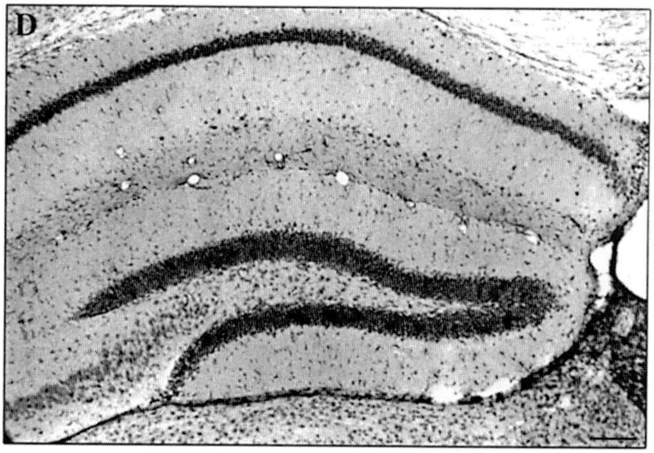

6D

The cellular localization of leptin-ir within brain cells should offer insight into the putative role of neural leptin. For example, as noted above, studies are needed to determine whether labeled cells are neuronal and/or glial. In some cells, leptin-ir is confined to the nucleus whereas in others (e.g. mouse SCN and median eminence; fig. 6C) it appears to be cytoplasmic. Some of the cytoplasmic staining, such as in median eminence or in tanycytes (fig. 6B), might reflect uptake of leptin from CSF or blood whereas in the SCN this is unlikely.

By analogy with other nuclear-localized signalling molecules (e.g. estradiol; FOS) we suggest that leptin may regulate gene expression in this way.

Such a mechanism is observed in the basal hypothalamus where leptin decreases gene expression for galanin, POMC, NPY and melanin concentrating hormone [Sahu, 1998; Kalra et al., 1999]. At present there is no evidence for this outside of the hypothalamus. However, a more global influence of leptin in brain can be deduced from the leptin-deficient ob/ob mouse. These mice are known to have unusually small brains, with decreased neuronal size, reduced myelination and low DNA levels [Kozak, 1971; Bereiter and Jeanrenaud, 1979; Rossier et al., 1979; van der Kroon and Speijers, 1979]. Leptin-insensitive db/db mice also have significantly smaller brains than control mice, accompanied by lower neuronal density and function [Garris and Williams, 1984; Timmers et al., 1990; Vannucci et al., 1997]. Two recent studies indicate that leptin regulates brain development and the expression of neuronal and glial proteins [Steppan and Swick, 1999; Ahima et al., 1999]. Treatment of ob/ob mice (age: 4 weeks) with leptin, daily for 2 weeks, reverses the decrease in brain size and increases total brain DNA [Steppan and Swick, 1999]. A more detailed examination of the effects of leptin replacement revealed an increase in brain weight and protein content, and a restoration of levels of growth-associated protein, syntaxin-1 and synaptosomal associated protein-25 [Ahima et al., 1999]. Note that these changes were not confined to hypothalamus but were also seen in neocortex. The data are suggestive of an important role for leptin in brain development. It is noteworthy that Ahima et al. [1999] also demonstrated the occurrence of neurodegeneration in frontal cortex and hippocampus of untreated ob/ob and db/db mice as compared to wild-type controls. However this was not corrected by treatment with peripherally administered lepin.

In summary, our findings that leptin gene expression and leptin protein are detectable in brain complements the suggestion that leptin may be an important factor in brain development [Ahima et al., 1999; Steppan and Swick, 1999]. The complex interplay of leptin with circulating hormones such as glucocorticoids and thyroid hormone, which also regulate brain maturation, will necessitate in vitro studies on the putative trophic influences of leptin.

An additional role for brain leptin could be as a neurotransmitter/neuromodulator acting at receptors in the cell membrane. There is evidence for rapid, nongenomic effects of leptin on synaptic transmission studied in hypothalamic slice preparations in vitro [Glaum et al., 1996; Powis et al., 1998; Sasaki et al., 1998]. Subnanomolar concentrations of leptin were rapidly (<1 min) and reversibly inhibitory tested on ARC neurons, but had little effect on neurons obtained from Zucker rats in which mutated leptin receptors are nonfunctional. A somewhat slower (5–15 min) hyperpolarizing effect has also been reported in hypothalamic glucose-sensitive neurons from lean rats [Spanswick et al., 1997], but not in tissue from Zucker rats. Depolarizing effects of leptin have been identified in neurons of the hypothalamic paraventricular nucleus [Powis et al., 1998]. Thus, all of these synaptic effects are clearly dependent on the presence of functional OBR.

Little work of this type has been done on extra-hypothalamic regions, which also possess high levels of leptin receptors. Lee and Morris [1998] observed an inhibitory effect of leptin on depolarization-induced NPY release from hypothalamus, cerebral cortex and medulla oblongata. No effects were seen in tissue from Zucker rats. Very small concentrations of leptin (10^{-14} and 10^{-12} M) have been shown to facilitate induction of LTP in rat hippocampal slices [Oomura et al., 1998]. Both these studies indicate that leptin receptors in non-hypothalamic areas are able to regulate neural activity. More importantly, the effect of leptin on LTP raises the possibility that leptin is involved in synaptic events unrelated to food intake and body weight control.

Summary and Suggestions for Future Progress

We have summarized the evidence that leptin has a dual role as a neurotransmitter/neuromodulator and as a trophic influence in the developing brain. Much of this information is already in the literature and we have reinterpreted some of the conclusions in light of our demonstration that leptin gene expression is readily detected in rat brain. Others have suggested such a possibility. For example, Wiesner et al. [1999] observed higher levels of leptin in human venous plasma exiting the brain compared to arterial levels. They hypothesized that the brain is a source of leptin. Our data confirm that this is correct. Also, in a study on hematopoiesis, Bennett et al. [1996] mention without comment the detection of leptin mRNA by RT-PCR in fetal mouse brain. In a recent editorial, Reichlin [1999], in a studied approach, also speculated that leptin may be a secretion of the brain.

There are other important issues concerning the expression of leptin in the brain. Detection of brain ob mRNA in species other than rat is necessary to establish a significant role for leptin in brain function. Identification of the cell types in which leptin is expressed is crucial for the understanding of neural leptin pathways. Our observation that leptin appears to be bound in cell nuclei provides another level of complexity to an already rich array of membrane-bound leptin receptors [Tartaglia, 1997; Fei et al., 1997]. Finally, leptin might influence brain development through a nuclear targeting system [Jans and Hassan, 1998].

Acknowledgements

This work was funded by grants from the Dalhousie University Internal Medicine Research Foundation and the QEII Research Foundation (to E.U.), the IWK Grace Research Foundation and the Medical Research Council of Canada (M.W.).

We are grateful to Diane Wilkinson and Cindee Leopold for invaluable technical assistance and to our collaborators Dr. Paul Murphy and Dr. Audrey Li for their generous support.

References

Ahima RS, Bjorbaek C, Osei S, Flier JS: Regulation of neuronal and glial proteins by leptin: Implications for brain development. Endocrinology 1999;140:2755–2762.

Bado A, Levasseur S, Attoub S, Kermorgant S, Laigneau JP, Bortoluzzi MN, Moizo L, Lehy T, Guerre-Millo M, Le Marchand-Brustel Y, Lewin MJM: The stomach is a source of leptin. Nature 1998; 394:790–793.

Banks WA, Kastin AJ, Huang W, Jaspan JB, Maness LM: Leptin enters the brain by a saturable system independent of insulin. Peptides 1996;17:305–311.

Baskin DG, Schwartz MW, Seeley RJ, Woods SC, Porte JD, Breininger JF, Jonak Z, Schaefer J, Krouse M, Burghardt C, Campfield LA, Burn P, Kochan JP: Leptin receptor long-form splice-variant protein expression in neuron cell bodies of the brain and co-localization with neuropeptide Y mRNA in the arcuate nucleus. J Histochem Cytochem 1999;47:353–362.

Bennett BD, Solar GP, Yuan JQ, Mathias J, Thomas GR, Matthews W: A role for leptin and its cognate receptor in hematopoiesis. Curr Biol 1996;6:1170–1180.

Bereiter DA, Jeanrenaud B: Altered neuroanatomical organization in the central nervous system of the genetically obese (ob/ob) mouse. Brain Res 1979;165:249–260.

Bjørbaek C, Uotani S, da Silva B, Flier JS: Divergent signaling capacities of the long and short isoforms of the leptin receptor. J Biol Chem 1997;272:32686–32695.

Bjørbaek C, Elmquist JK, Michl P, Ahima RS, van Bueren A, McCall AL, Flier JS: Expression of leptin receptor isoforms in rat brain microvessels. Endocrinology, 1998;139:3485–3491.

Bottner A, Scherbaum WA, Chrousos GP, Bornstein SR: Leptin expression in the mouse adrenal gland. Endocrine Society 81st Meeting 1999, abstr P2-404.

Bouchaud C, Bosler O: The circumventricular organs of the mammalian brain with special reference to monoaminergic innervation. Int Rev Cytol 1986;105:283–327.

Bouloumie A, Drexler HC, Lafontan M, Busse R: Leptin, the product of Ob gene, promotes angiogenesis. Circulation Res 1998;83:1059–1066.

Broadwell RD, Sofroniew MV: Serum proteins bypass the blood-brain fluid barriers for extracellular entry to the central nervous system. Exp Neurol 1993;120:245–263.

Campfield LA, Smith FJ: Overview: Neurobiology of OB protein (leptin). Proc Nutr Soc 1998;57:429–440.

Dagogo-Jack S: Regulation and possible significance of leptin in humans: Leptin in health and disease. Diabetes Rev 1999;7:23–38.

Dallongeville J, Fruchart JC, Auwerx J: Leptin, a pleiotropic hormone: Physiology, pharmacology, and strategies for discovery of leptin modulators. J Med Chem 1998;41:5337–5352.

De Matteis R, Cinti S: Ultrastructural immunolocalization of leptin receptor in mouse brain. Neuroendocrinology 1998;68:412–419.

Devaskar SU, Giddings SJ, Rajakumar PA, Carnaghi LR, Menon RK, Zham DS: Insulin gene expression and insulin synthesis in mammalian neuronal cells. J Biol Chem 1994;269:8445–8454.

Diano S, Kalra SP, Horvath TL: Leptin receptor immunoreactivity is associated with the golgi apparatus of hypothalamic neurones and glial cells. J Neuroendocrinol 1998;10:647–650.

Diano S, Kalra SP, Sakamoto H, Horvath TL: Leptin receptors in estrogen receptor-containing neurons of the female rat hypothalamus. Brain Res 1998;812:256–259.

Elmquist JK, Ahima RS, Maratos-Flier E, Flier JS, Sapier CB: Leptin activates neurons in ventrobasal hypothalamus and brainstem. Endocrinology 1997;138:839–842.
Elmquist JK, Scammell TE, Saper CB: Mechanisms of CNS response to systemic immune challenge: The febrile response. Trends Neurosci 1997;20:565–570.
Elmquist JK, Bjørbaek C, Ahima RS, Flier JS, Saper CB: Distributions of leptin receptor mRNA isoforms in the rat brain. J Comp Neurol 1998;395:535–547.
Elmquist JK, Maratos-Flier E, Saper CB, Flier JS: Unraveling the central nervous system pathways underlying responses to leptin. Nature Neurosci 1998;1:445–450.
Elmquist JK, Elias CF, Saper CB: From lesions to leptin: Hypothalamic control of food intake and body weight. Neuron 1999;22:221–232.
Fei H, Okano HJ, Li C, Lee GH, Zhao C, Darnell R, Friedman JM: Anatomic localization of alternatively spliced leptin receptors (Ob-R) in mouse brain and other tissues. Proc Natl Acad Sci USA 1997; 94:7001–7005.
Finn PD, Cunningham MJ, Pau KYF, Spies HG, Clifton DK, Steiner RA: The stimulatory effect of leptin on the neuroendocrine reproductive axis of the monkey. Endocrinology 1998;139:4652–4662.
Friedman JM, Halaas L: Leptin and the regulation of body weight in mammals. Nature 1998;395:763–770.
Garris DR, Williams SK, Coleman DL, Morgan CR: Glucose utilization by the mouse brain: Influence of age and diabetes. Brain Res 1984;317:141–146.
Glaum SR, Hara M, Bindokas VP, Lee CC, Polonsky KS, Bell GI, Miller RJ: Leptin, the obese gene product, rapidly modulates synaptic transmission in the hypothalamus. Mol Pharmacol 1996;50: 230–235.
Golden PL, Maccagnan TJ, Pardridge WM: Human blood-brain barrier leptin receptor. Binding and endocytosis in isolated human brain microvessels. J Clin Invest 1997;99:14–18.
Goldstein GW, Betz AL: The blood-brain barrier. Sci Am 1986;255:74–83.
Guan XM, Hess JF, Yu H, Hey PJ, van der Ploeg LHT: Differential expression of mRNA for leptin receptor isoforms in the rat brain. Mol Cell Endocrinol 1997;133:1–7.
Håkansson ML, Brown H, Ghilardi N, Skoda RC, Meister B: Leptin receptor immunoreactivity in chemically defined target neurons of the hypothalamus. J Neurosci 1998;18:559–572.
Håkansson ML, Meister B: Transcription factor STAT3 in leptin target neurons of rat hypothalamus. Neuroendocrinology 1998;68:420–427.
Herkenham M, Lee HY, Baker RA: Temporal and spatial patterns of c-fos mRNA induced by intravenous interleukin-1: A cascade of non-neuronal cellular activation at the blood-brain barrier. J Comp Neur 1998;400:175–196.
Hoggard N, Mercer JG, Rayner DV, Moar K, Trayhurn P, Williams LM: Localization of leptin receptor mRNA splice variants in murine peripheral tissues by RT-PCR and in situ hybridization. Biochem Biophys Res Commun 1997;232:383–387.
Hughes P, Dragunow M: Induction of immediate-early genes and the control of neurotransmitter-regulated gene expression within the nervous system. Pharmacol Rev 1995;47:133–178.
Inui A: Feeding and body-weight regulation by hypothalamic neuropeptides – Mediation of the actions of leptin. Trends Neurosci 1999;22:62–67.
Jans DA, Hassan G: Nuclear targeting by growth factors, cytokines, and their receptors: A role in signaling? BioEssays 1998;20:400–411.
Kalra SP, Dube MG, Pu S, Xu B, Horvath TL, Kalra PS: Interacting appetite-regulating pathways in the hypothalamic regulation of body weight. Endocrine Rev 1999;20:68–100.
Kielar D, Clark JSC, Ciechanowicz A, Kurzawski G, Sulikowski T, Naruszewicz M: Leptin receptor isoforms expressed in human adipose tissue. Metabolism 1998;47:844–847.
Kozak L: Neurochemical correlates of obesity. 43rd Annual Report Jackson Lab 1971:83.
Lee J, Morris MJ: Modulation of neuropeptide Y overflow by leptin in the rat hypothalamus, cerebral cortex and medulla. NeuroReport 1998;9:1575–1580.
Li HY, Wang LL, Yeh RS: Leptin immunoreactivity in the central nervous system in normal and diabetic rats. NeuroReport 1999;10:437–442.
Malik KF, Young III WS: Localization of binding sites in the central nervous system for leptin (OB protein) in normal, obese (ob/ob), and diabetic (db/db) C57BL/6J mice. Endocrinology 1996;137: 1497–1500.

Maness LM, Kastin AJ, Farrell CL, Banks WA: Fate of leptin after intracerebroventricular injection into the mouse brain. Endocrinology 1998;139:4556–4562.

Mansour A, Fox CA, Akil H, Watson SJ: Opioid-receptor mRNA expression in the rat CNS: Anatomical and functional implications. Trends Neurosci 1995;18:22–29.

Mantzoros CS, Moschos SJ: Leptin: In search of role(s) in human physiology and pathophysiology. Clin Endocrinol 1998;49:551–567.

Mantzoros CS: The role of leptin in human obesity and disease: A review of current evidence. Ann Intern Med 1999;130:671–680.

Meisenberg G, Simmons WH: Peptides and the blood-brain barrier. Life Sci 1983;32:2611–2623.

Morash B, Li A, Murphy P, Wilkinson M, Ur E: Leptin gene expression in the brain and pituitary gland. Endocrinology 1999; in press.

Natarajan M, Wilkinson M: Recovery of hypothalamic NMDA-induced c-fos expression following neonatal glutamate (MSG) lesions. Dev Brain Res 1997;102:97–104.

O'Brien SN, Welter BH, Price TM: The presence of leptin in breast tumors, breast cancer cell lines and a normal breast epithelial cell line. Endocrine Society 81st Meeting 1999;abstr P3-200.

Oomura Y, Sasaki K, Hori N, Shiraishi T, Takeda H, Tsuji H, Matsumiya T: Effects of leptin on passive avoidance and morris water maze tasks and on hippocampal activity in rats. Soc Neurosci Abstracts 1998;24:abstr 174.6.

Porte D Jr, Seeley RJ, Woods SC, Baskin DG, Figlewicz DP, Schwartz MW: Obesity, diabetes and the central nervous system. Diabetologia 1998;41:863–881.

Powis JE, Bains JS, Ferguson AV: Leptin depolarizes rat hypothalamic paraventricular nucleus neurons. Am J Physiol 1998;274:R1468–R1472.

Ramsey JJ, Kemnitz JW, Colman RJ, Cunningham D, Swick AG: Different central and peripheral responses to leptin in rhesus monkeys: Brain transport may be limited. J Clin Endocrinol Metab 1998;83:3230–3235.

Reichlin S: Editorial: Is leptin a secretion of the brain? J Clin Endocrinol Metab 1999;84:2267–2269.

Rossier J, Rogers J, Shibasaki T, Guillemin R, Bloom FE: Opioid peptides and α-melanocyte-stimulating hormone in genetically obese (ob/ob) mice during development. Proc Natl Acad Sci USA 1979;76:2077–2080.

Sahu A: Evidence suggesting that galanin (GAL), melanin-concentrating hormone (MCH), neurotensin (NT), proopiomelanocortin (POMC) and neuropeptide Y (NPY) are targets of leptin signaling in the hypothalamus. Endocrinology 1998;139:795–798.

Sasaki K, Nagamori K, Shiraishi T, Muramoto K, Oomura Y: Effects of leptin and neuropeptide-Y (NPY) on rat arcuate neurons. Soc Neurosci Abstr 1998;24:abstr 175.12.

Sawchenko PE: Toward a new neurobiology of energy balance, appetite, and obesity. The anatomists weigh in. J Comp Neurol 1998;402:435–441.

Schechter R, Yanovitch T, Abboud M, Johnson G III, Gaskins J: Effects of brain endogenous insulin on neurofilament and MAPK in fetal rat neuron cell cultures. Brain Res 1998;808:270–278.

Seeley RJ, Yagaloff KA, Fisher SL, Burn P, Thiele TE, van Dijk G, Baskin DG, Schwartz MW. Melanocortin receptors in leptin effects. Nature 1997;390:349.

Shioda S, Funahashi H, Nakajo S, Yada T, Maruta O, Nakai Y: Immunohistochemical localization of leptin receptor in the rat brain. Neurosci Lett 1998;243:41–44.

Sierra-Honigmann MR, Nath AK, Murakami C, Garcia-Cardena G, Papapetropoulos A, Sessa WC, Madge LA, Schechner JS, Schwabb MB, Polverini PJ, Flores-Riveros JR: Biological action of leptin as an angiogenic factor. Science 1998;281:1683–1686.

Sinha MK, Sturis J, Ohannesian J, Magosin S, Stephens T, Heiman ML, Polonsky KS, Caro JF: Ultradian oscillations of leptin secretion in humans. Biochem Biophys Res Commun 1996a;228:733–738.

Sinha MK, Opentanova I, Ohannesian JP, Kolaczynski JW, Heiman ML, Hale J, Becker GW, Bowsher RR, Stephens TW, Caro JF: Evidence of free and bound leptin in human circulation. J Clin Invest 1996b;98:1277–1282.

Spanswick D, Smith MA, Groppi VE, Logan SD, Ashford MLJ: Leptin inhibits hypothalamic neurons by activation of ATP-sensitive potassium channels. Nature 1997;390:521–525.

Steppan CM, Swick AG: A role for leptin in brain development. Biochem Biophy Res Commun 1999;256:600–602.

Tang-Christensen M, Havel PJ, Jacobs RR, Larsen PJ, Cameron JL: Central administration of leptin inhibits food intake and activates the sympathetic nervous system in rhesus macaques. J Clin Endocrinol Metab 1999;84:711–717.
Tartaglia LA: The leptin receptor. J Biol Chem 1997;272:6093–6096.
Timmers KI, Palkovits M, Coleman DL: Unique alterations of neuropeptide content in median eminence, amygdala, and dorsal vagal complex of 3- and 6-week-old diabetes mutant mice. Metabolism 1990; 39:1158–1166.
Uotani S, Bjorbaek C, Tornoe J, Flier JS: Functional properties of leptin receptor isoforms internalization and degradation of leptin and ligand-induced receptor downregulation. Diabetes 1999;48:279–286.
Ur E, Morash B, Li A, Murphy PR, Wilkinson M: Leptin gene expression in the rat pituitary. Endocrine Society 81st Meeting 1999;abstr P3-91.
Vaisse C, Halaas JL, Horvath CM, Darnell JE, Stoffel M, Friedman JM: Leptin activation of Stat3 in the hypothalamus of wild-type and ob/ob mice but not db/db mice. Nature Genet 1996;14:95–97.
van der Kroon PHW, Speijers GJA: Brain deviations in adult obese-hyperglycemic mice (ob/ob). Metabolism 1979;28:1–3.
Van Dijk G, Thiele TE, Donahey JCK, Campfield LA, Smith FJ, Burn P, Bernstein IL, Woods SC, Seeley RJ: Central infusions of leptin and GLP-1-(7-36) amide differentially stimulate c-FLI in the rat brain. Am J Physiol 1996;271:R1096–R1100.
Van Heek M, Mullins DE, Wirth MA, Graziano MP, Fawzi AB, Compton DS, France CF, Hoos LM, Casale RL, Sybertz EJ, Strader CD, Davis HR Jr: The relationship of tissue localization, distribution and turnover to feeding after intraperitoneal ^{125}I-leptin administration to ob/ob and db/db mice. Horm Metab Res 1996;28:653–658.
Vannucci SJ, Gibbs EM, Simpson IA: Glucose utilization and glucose transporter proteins GLUT-1 and GLUT-3 in brains of diabetic (db/db) mice. Am J Physiol 1997;272:E267–E274.
Wang L, Martinez V, Barrachina MD, Taché Y: Fos expression in the brain induced by peripheral injection of CCK or leptin plus CCK in fasted lean mice. Brain Res 1998;791:157–166.
Wiesner G, Vaz M, Collier G, Seals D, Kaye D, Jennings G, Lambert G, Wilkinson D, Esler M: Leptin is released from the human brain: Influence of adiposity and gender. J Clin Endocr Metab 1999; 84:2270–2274.
Woods AJ, Stock MJ: Leptin activation in hypothalamus. Nature 1996;381:745.
Woods SC, Seeley RJ, Porte D Jr, Schwartz MW: Signals that regulate food intake and energy homeostasis. Science 1998;280:1378–1383.
Zhang Y, Proenca R, Maffei M, Barone M, Leopold L, Friedman JM: Positional cloning of the mouse obese gene and its human homologue. Nature 1994;372:425–432.

Subject Index